Advances in Intelligent Systems and Computing

Volume 653

Series editor

Janusz Kacprzyk, Polish Academy of Sciences, Warsaw, Poland
e-mail: kacprzyk@ibspan.waw.pl

About this Series

The series "Advances in Intelligent Systems and Computing" contains publications on theory, applications, and design methods of Intelligent Systems and Intelligent Computing. Virtually all disciplines such as engineering, natural sciences, computer and information science, ICT, economics, business, e-commerce, environment, healthcare, life science are covered. The list of topics spans all the areas of modern intelligent systems and computing.

The publications within "Advances in Intelligent Systems and Computing" are primarily textbooks and proceedings of important conferences, symposia and congresses. They cover significant recent developments in the field, both of a foundational and applicable character. An important characteristic feature of the series is the short publication time and world-wide distribution. This permits a rapid and broad dissemination of research results.

More information about this series at http://www.springer.com/series/11156

A.K. Saini · A.K. Nayak
Ram Krishna Vyas
Editors

ICT Based Innovations

Proceedings of CSI 2015

Springer

Editors
A.K. Saini
USMS, GGSIP University
New Delhi, Delhi
India

Ram Krishna Vyas
Institute of Life Long Learning (ILLL)
New Delhi, Delhi
India

A.K. Nayak
Indian Institute of Business Management
Patna, Bihar
India

ISSN 2194-5357 ISSN 2194-5365 (electronic)
Advances in Intelligent Systems and Computing
ISBN 978-981-10-6601-6 ISBN 978-981-10-6602-3 (eBook)
https://doi.org/10.1007/978-981-10-6602-3

Library of Congress Control Number: 2017952511

Printed on acid-free paper

This Springer imprint is published by Springer Nature
The registered company is Springer Nature Singapore Pte Ltd.
The registered company address is: 152 Beach Road, #21-01/04 Gateway East, Singapore 189721, Singapore

Preface

The last decade has witnessed remarkable changes in IT industry, virtually in all domains. The 50th Annual Convention, CSI-2015 on the theme "Digital Life" was organized as a part of CSI@50, by CSI at Delhi, the National Capital of the Country, during December 02–05, 2015. Its concept was formed with an objective to keep ICT community abreast of emerging paradigms in the areas of computing technologies and more importantly looking at its impact on the society.

Information and Communication Technologies (ICT) comprises of three main components; infrastructure, services, and product. These components include the Internet, infrastructure based/infrastructure less wireless networks, mobile terminals, and other communication medium. ICT is gaining popularity due to rapid growth in communication capabilities for real-time-based applications. New user requirements and services entail mechanisms for enabling systems to intelligently process speech and language-based input from human users. CSI-2015 attracted over 1500 papers from researchers and practitioners from academia, industry and government agencies, from all over of the world, thereby making the job of the Programme Committee extremely difficult. After a series of tough review exercises by a team of over 700 experts, 565 papers were accepted for presentation in CSI-2015 during the 3 days of the convention under ten parallel tracks. The Programme Committee, in consultation with Springer, the world's largest publisher of scientific documents, decided to publish the Proceedings of the presented papers, after the convention, in 10 topical volumes, under ASIC series of the Springer, as detailed hereunder:-

1. Volume # 1: ICT based Innovations
2. Volume # 2: Next Generation Networks
3. Volume # 3: Nature Inspired Computing
4. Volume # 4: Speech and Language Processing for Human-Machine Communications
5. Volume # 5: Sensors and Image Processing
6. Volume # 6: Big Data Analytics

7. Volume # 7: Systems and Architecture
8. Volume # 8: Cyber Security
9. Volume # 9: Software Engineering
10. Volume # 10: Silicon Photonics and High Performance Computing

We are pleased to present before you the proceedings of the Volume # 1 on "ICT based Innovations". The idea encompasses innovations in health care, education, e-governance as well as various fields of human endeavor. It highlights how Information and Communication Technologies (ICTs) are the key players in improving governance. The title also brings out how ICT can contribute in capacity building through digital systems.

Information and Communications Technologies (ICT) is an umbrella term that includes computing and communication devices and systems like e-learning, e-governance, e-banking, etc. The ICT dimensions include any system that stores, retrieves, manipulates, transmit, or receive information electronically in a digital form, e.g., personal computers, digital television, email, robots, etc. ICT has the potential to bridge the digital divide among various sections of the society. It discusses various services and applications associated with ICTs including special services such as videoconferencing and e-learning.

The title "ICT Based Innovations" aims at covering the wide spectrum of applications of ICT based technologies in different domains. The volume includes scientific, original, and high-quality papers presenting novel research, ideas, and explorations of new vistas in speech and language processing like speech recognition, text recognition, embedded platform for information retrieval, segmentation, filtering and classification of data, emotion recognition, etc. The aim of this volume is to provide a stimulating forum for sharing knowledge and results in model, methodology, and implementations of speech and language processing tools. Its authors are researchers and experts of these domains. This volume is designed to bring together researchers and practitioners from academia and industry to focus on extending the understanding and establishing new collaborations in these areas. It is the outcome of the hard work of the editorial team, who have relentlessly worked with the authors and steered up the same to compile this volume. It will be a useful source of reference for the future researchers in this domain. Under the CSI-2015 umbrella, we received over 200 papers for this volume, out of which 25 papers are being published, after rigorous review processes, carried out in multiple cycles.

On behalf of organizing team, it is a matter of great pleasure that CSI-2015 has received an overwhelming response from various professionals from across the country. The organizers of CSI-2015 are thankful to the members of *Advisory Committee, Programme Committee and Organizing Committee* for their all-round guidance, encouragement, and continuous support. We express our sincere gratitude to the learned *Key note Speakers* for support and help extended to make this event a grand success. Our sincere thanks are also due to our *Review Committee Members* and the *Editorial Board* for their untiring efforts in reviewing the manuscripts, giving suggestions and valuable inputs for shaping this volume.

We hope that all the participants/delegates will be benefitted academically and wish them all the best for their future endeavors.

We also take the opportunity to thank the entire team from Springer, who have worked tirelessly and made the publication of the volume a reality. Last but not the least, we thank the team from Bharati Vidyapeeth's Institute of Computer Applications and Management (BVICAM), New Delhi, for their untiring support, without which the compilation of this huge volume would not have been possible.

New Delhi, India A.K. Saini
Patna, India A.K. Nayak
New Delhi, India Ram Krishna Vyas
March 2017

The Organization of CSI-2015

Chief Patron

Padmashree Dr. R. Chidambaram, *Principal Scientific Advisor, Government of India*

Patrons

Prof. S.V. Raghavan, Department of Computer Science, IIT Madras, Chennai

Prof. Ashutosh Sharma, Secretary, Department of Science and Technology, Ministry of Science of Technology, Government of India

Chair, Programme Committee

Prof. K.K. Aggarwal, Founder Vice Chancellor, GGSIP University, New Delhi

Secretary, Programme Committee

Prof. M.N. Hoda, Director, Bharati Vidyapeeth's Institute of Computer Applications and Management (BVICAM), New Delhi

Advisory Committee

Editorial Board

Contents

About the Editors

Prof. A.K. Saini is a Professor of Information Technology Management. He obtained his doctoral degree from the Faculty of Management Studies (FMS) at Delhi University, India. Before joining the University School of Management Studies (USMS), Guru Gobind Singh (GGS) Indraprastha University, he worked with the FMS, University of Delhi. He has over 30 years industrial and academic experience, including 6 years in the industry. Further, Dr. Saini has published eight books and over 80 papers in national and international journals/conferences. He has traveled extensively and has visited Europe under the Teacher-Exchange Programme for teaching courses abroad. He has been proactively involved with professional associations and is a Senior Life Member of the Computer Society of India (CSI), Fellow of the Institution of Electronics and Telecommunication Engineers (IETE), and Life Member of the All India Management Association (AIMA) and Operational Research Society of India (ORSI). Dr. Saini serves on the Board of Studies of several State and Central Universities and Institutions. His major interests include: Computer Applications, Information Systems, Databases, Technology and Innovation Management, Knowledge Management, and Healthcare Systems.

Prof. A.K. Nayak is Director of the Indian Institute of Business Management, Patna. He has over 30 years of experience in teaching, training, research, and administration in the fields of Computer Science and Information Technology (IT). He served as the President of the Information and Communications Technologies (ICT Section), Indian Science Congress Association; Zonal Coordinator, National Institute of Electronics and Information Technology, Government of India; and National Chairman, Data Security Applications. At present, he is the Chairman, Bihar-Jharkhand Section of the Indian Society for Technical Education and Chairman of the Publication Committee of the Computer Society of India (CSI). He is actively associated with several professional societies and has authored several books and technical papers in the areas of Computer Science and Information Technology.

Prof. Ram Krishna Vyas has over 30 years of experience in the field of Information Technology (IT). Presently, he is working at the Institute of Life Long Learning (ILLL), University of Delhi, India. He is actively involved in coordinating the Massive Open Online Courses (MOOCs) and National Mission on Education through Information and Communication Technology (NMEICT) project, sponsored by the Ministry of Human Resource Development (MHRD) for E-content Development, on behalf of the University of Delhi.

FPKIVS—A Stellar Approach to Voting Systems in India

Digvijay Khojare, Vaibhav Chaudhary, Malvika Malviya and Shweta Shukla

Abstract This paper introduces system FPKIVS (Finger Print Kerberos Internet Voting System) based on i-voting which removes the geographic constraints of voting from a hometown. This paper illustrates the modern way of voting. It reduces the shortcomings of conventional voting system. The key element of our system is fingerprint; any voter will be able to cast his vote through his fingerprint. With the help of simple machinery that is a computer system with secured connections and a fingerprint scanner, our system will serve as a constructive way to vote from any place by going to the nearest booth. We have linked fingerprints with the Voter ID. Thus, our system will be a perfect alternative to the traditional voting system.

Keywords Kerberos · FPKIVS · Voting server (VS) · Authentication server (AS) · Voter authentication server (VAS)

1 Introduction

In a country like India, which is having world's largest democracy, Voting is an essential activity. In the current traditional system, many flaws have been observed like many frauds and bogus voting takes place. Also, a voter has to go to his/her respective hometown, which is a constraint for the ones who work in other cities. Thus, we need a mechanism which will enable the user to vote from anywhere. In case of the traditional voting system, a voter is bound to vote from his own city to which he belongs. This is not feasible for the ones living in other cities. Our system makes the users vote from anywhere. It assures security which is the most important thing to be looked after. In case of handicapped people, the normal voter ID cards can be used for the process.

D. Khojare (✉) · V. Chaudhary · M. Malviya · S. Shukla
P.R.Patil College of Engineering and Technology Amravati, Amravati, India
e-mail: digvijaykhojare@gmail.com

A.K. Saini et al. (eds.), *ICT Based Innovations*, Advances in Intelligent Systems and Computing 653, https://doi.org/10.1007/978-981-10-6602-3_1

1.1 Objectives

The main objectives of this study are

- To develop a general prototype system that provides security and trusted electronic voting.
- To remove the geographic constraints of voting from a home town.

1.2 Scope of Study

This project tries to remove the geographic constraints using fingerprints.

2 Literature Review

Tadayoshi Kohno and Adam Stubblefield proposed that a voting system must be comprehensible and used by the entire voting population, irrespective of age, infirmity, or disability. It is necessary to provide access to such a diverse population and with strong security, e-voting could be a great change over current system. The code should have security relevant flaws, or by malicious insiders like election officials, the developers of the voting system, and the developers of the embedded operating system on which the voting system runs. If any political party generates error codes into the voting system software, then the advantage of pre-existed flaws will lead to improper results of the election. So by avoiding the above errors, a strongly secured system e-voting can well be implemented [1].

Punch Scan is a cryptographic voting system that is easily used by the voters as well as by election officials, which provides transparent and reliable process. It will also provide public verifiability and election integrity. Also, voter privacy will be enhanced [2].

Hussein Khalid Abd-alrazzq1, Mohammad S. Ibrahim & Omar Abdulrahman Dawood proposed that the i-voting (Internet Voting) system is a practicable alternative. Voters can vote in elections over the Internet, solving the problems of geographical restrictions and thus increasing the rate of voting. The main goal of a secure i-voting system is to ensure the privacy and accuracy of votes. Also, the problems of bogus voting will be eliminated [3].

A. Hazimeh, Mutaz KH. Khazaaleh, Khairall M. Al-Talafha proposed that contrarily to the traditional way of voting, the voting system is important because the polling tasks can be performed electronically without sacrificing voter privacy and completely avoiding the frauds. To make sure that the system performs these tasks well, it is useful to develop a set of criteria for evaluation of system performance. The criteria are accuracy, democracy, convenience, flexibility, privacy, verifiability and mobility [4].

Trisha Patel, Maitri Chokshi Nikhil Shah have pointed out the flaws as follows: Drawbacks with the existing voting system:

- Voter has to go to the ward where he\she was born or from where he had made the voter ID.
- Chances of dummy voting are more because if authorised person is not honest as he/she is required to be so, he might perform illegal task of voting for a particular party person. Due to this, malfunctioning chances of violence may raise or disturb the ongoing process. In few cases, voter is registered at more than one area so there are chances of vote recorded twice.
- Voter has different reasons to deny voting such as being physically sick or not being able to reach to the allocated voting booth.
- To carry out an election, a huge manpower is required and also the cost increases as the labour involvement increases; an estimation was that 6400 volunteers are required to fight for one Loksabha seat [5].

3 Proposed Work

We have proposed an i-voting system FPKIVS (Finger Print Kerberos Internet Voting System). The voting process will be carried out mainly in two phases namely, registration and voting. To ensure security, we will use Kerberos protocol. The two phases are illustrated below:

3.1 Registration

In registration phase, the interested electorate will go to the authorised registration camps and will provide the necessary information along with the required documents. The key element for registration will be the fingerprints. Each electorate has to scan his/her fingerprints which will be used as the voter ID as well as identity proof for the electorate.

We have also proposed a scheme which allows under age 18 people to register for voter ID at the time of registration for Aadhar card. Their voter ID will be generated automatically after they have reached the eligible age 18. Our system also provides the scheme for documents centralisation which will reduce the tedious task of registering for various other identity cards for which you will have to register only once for different documents like Aadhar card, Pan card, Voter ID, etc.

After the authentication of the applicant, the registration committee will generate the voter ID. The voter ID and the fingerprint will be linked together. In case of handicapped people, the voter ID can be used as an alternative to the fingerprints.

In the registration phase, the user will go to the authorised registration camps and will provide the necessary information as need. In the voting phase, the user will go

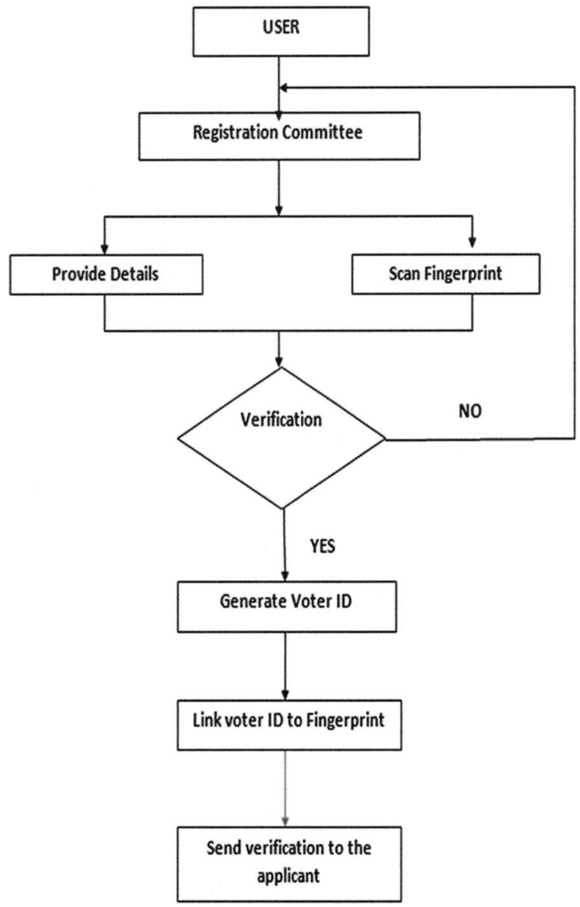

Fig. 1 Registration process

to any authorised voting booth. The booths will be equipped with a fingerprint scanner. The fingerprint scanner will provide the identity proof of the voter, which will directly display a voting page which will have candidates of his/her ward. The pictorial representation of registration process is shown in Fig. 1.

3.2 Voting

In this phase, the electorate will cast the vote. The process includes the following steps:

1. Visiting the nearby booth;
2. Scanning;
3. Authentication;

4. Cast vote;
5. Completion.

3.2.1 Visiting the Nearby Booth

On the elections day, the voter has to go to the nearby voting booth.

3.2.2 Scanning

In this step, the voter will have to scan the fingerprints which will be used to authenticate the voter.

3.2.3 Authentication

In this step, the identity of voter will be verified. If has successfully registered, a voting page will be displayed consisting of the candidates of his/her particular area (Fig. 2).

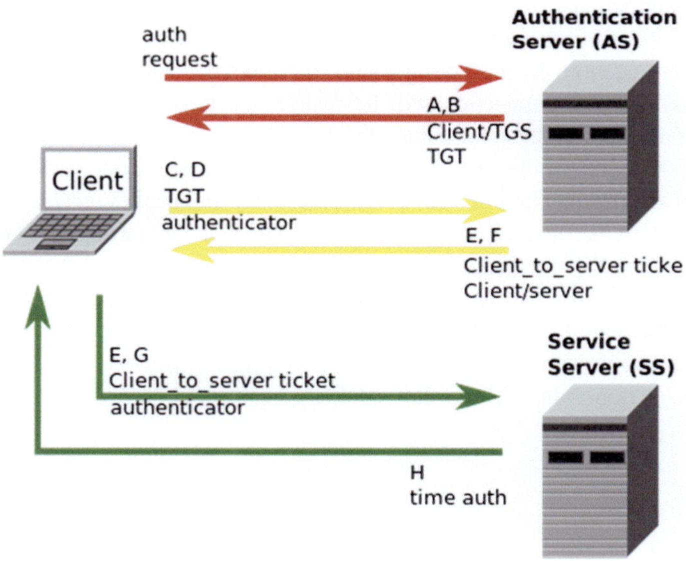

Fig. 2 Kerberos

3.2.4 Cast Vote

In this step, the user can cast the vote by simply clicking on the party symbol.

3.2.5 Completion

Once the voter has successfully casted the vote, the vote will be recorded in the centralised database by the voting server (VS), and the page will be excited. There will be a mechanism which will keep logs of the voters who have already casted the votes by incrementing a counter, say vote flag whose value will avoid the bogus votes. The voting process is illustrated in Fig. 3.

4 Network Security

To assure a secure network connection, we will be using Kerberos authentication protocol. Kerberos consists of a centralised authentication server which authenticates users to servers and servers to the users. We use the Kerberos with public key. Cryptography increases the security, authenticity and efficiency of the system. The client authentication will be the responsibility of the election commission. They will be responsible for generating the public and private keys for the server–client–server authentication. In pre-voting phase, the election commission will authenticate the server. The connection will be maintained throughout the voting phase, thus assuring the secure communication over the network.

The system can further be improved by using Elliptic Curve (EC) arithmetic operations to generate public and private keys for public key cryptography; the principles of EC compared to RSA or El Gamal offer the same security but with smaller bit size, thereby reducing processing overhead. The concept of blind signature can also be used which was first proposed by David Chaum in 1982, which blinds the message before sending thus making it untraceable to the attackers. In this way, we can assure the network security so that we can concentrate on making the databases non-vulnerable to the attackers.

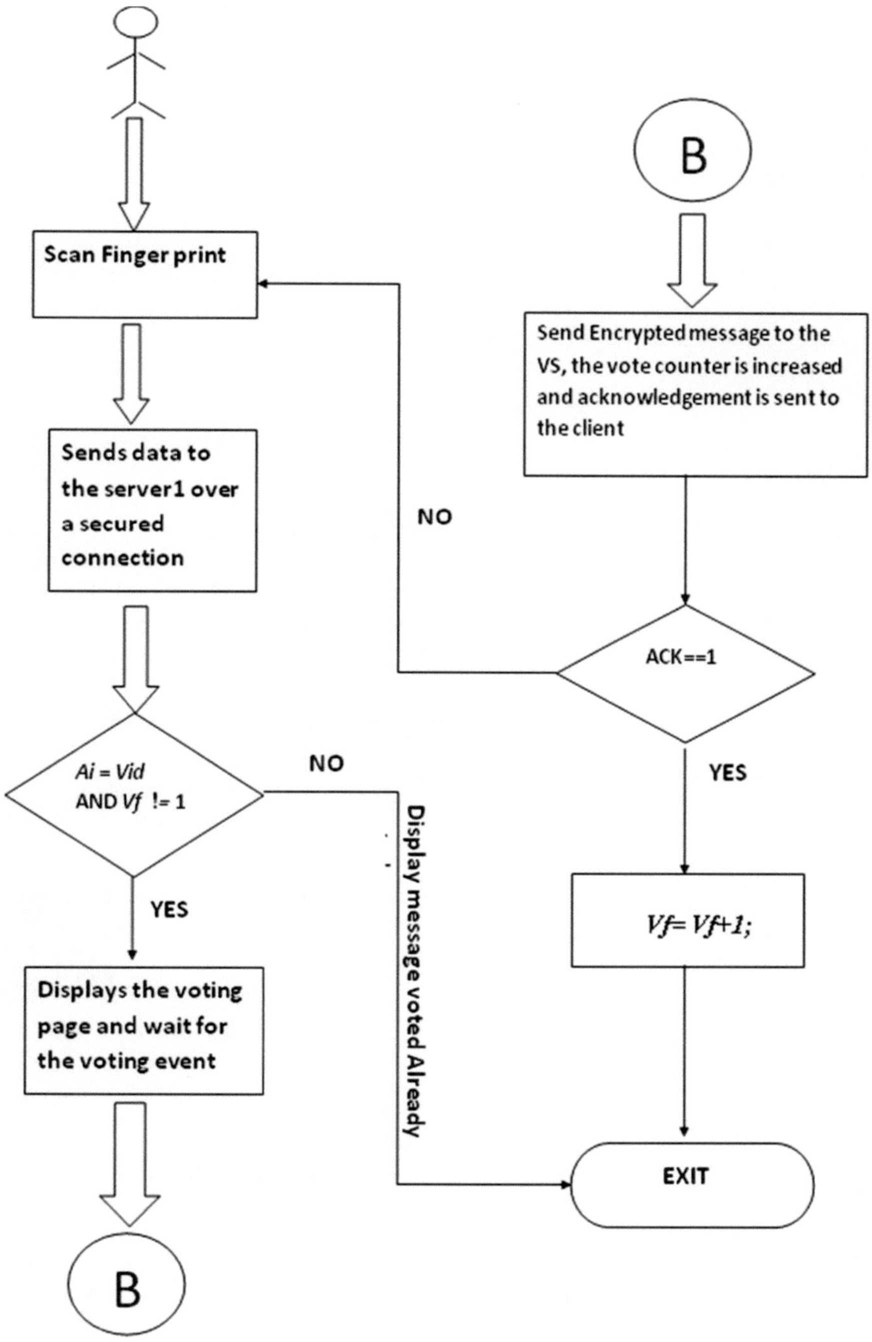

Fig. 3 Voting process

5 Proposed Algorithms

For Voting stage:

(1) The electorate scans the finger print.
    Ai = (***Fi***).
    For i = 1 to N … electorate population.
    For all ***Ai*….** Authenticate the electorate.
(2) For all ***Ai*,** check for the match in the database such that,
    If ***Ai*** = ***Vid*** AND ***Vf*** **! = 1** is true, authenticate the user
    Where ***Vid*** **=** Voter Id**,** ***Vf*** = voter flag**.**
    Else go to 8.
(3) Display the voting page ***Pi***.

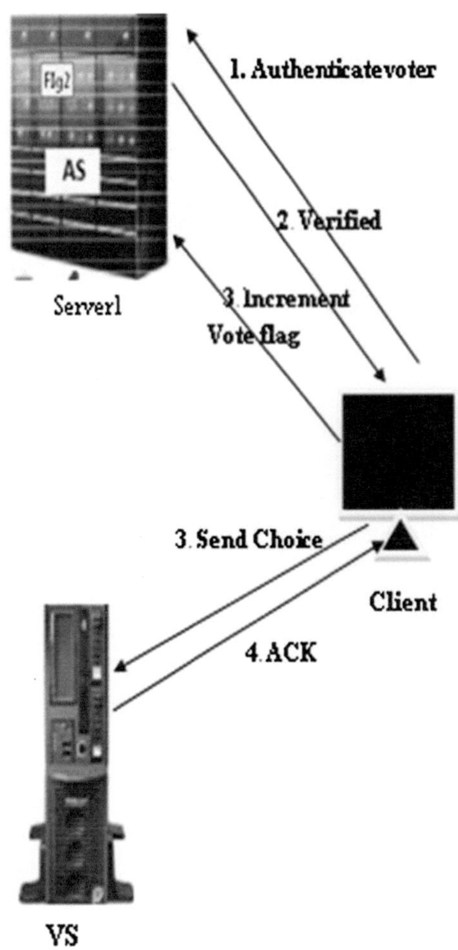

Fig. 4 Client–server communication

(4) `Wait for the voting event.`
(5) `Submit Event will generate an encrypted message and sends it to VS.`
(6) `VS Records vote and increases counter and send an acknowledgment to the client machine (voting booth.`
(7) `If` ***ACK = 1***`, increment the vote flag such that,`
`all` ***Ei***`,` *If* **ACk** `=1,` ***Vf*** `=` ***Vf****+1;*
Else go to 1.
(8) `Exit voting page.`

The overall phase of client–server communication is shown in Fig. 4.

6 Advantages of Our Proposed System

- Rate of voting will increase.
- A non-localite can vote from any nearest booth.
- If he has successfully registered, then there is no need to carry anything, as fingerprints can be used as identity cards.
- It will completely remove the fraud of bogus voters.
- Also corruption during the process will be minimised.
- Labour work will be reduced.
- It will reduce the tedious counting of votes, as results will be obtained instantly.
- More reliable.
- Secured using Kerberos protocol.
- Time saving, compared to the traditional system as people will not have to wait and stand in long queues.

7 Conclusion

This paper illustrates a proficient internet voting system provided with security protocol. Main goal of this system is that it will increase the percentage of voters especially in country like India. The vital key used in our system is the uniqueness of fingerprints. We all know in democracy form of governance voting is crucial for the country, which is restricted by geographic constraints. Because of centralised repository, data will be accessible at any moment of time and backup of the data will be possible too. It allows fast and updated result at that second. It will reduce the amount of budget sanctioned by the government as fewer resources are required. The prerequisite is that the database needs to be updated every year or before election so that new eligible citizens may register. It will completely eradicate concept of bogus voting as once fingerprint scanned will not be scanned again. Thus our proposed system will serve better than the current system.

Acknowledgements We would like to acknowledge Prof. Ajay B. Gadicha for his consistent guidelines and valuable support for publishing this paper. Also, we would like to thank our friend Miss. Pooja H. Lulla for her knowledge and support.

References

1. Kohno, T., Stubblefield, A., Rubin, A.D., Wallach D.S.: Analysis of an electronic voting system (2004, February 27)
2. Weldemariam, K., Villafiorita, A.: A survey: electronic voting development and trends. **3**(1) (2013, January). ISSN: 2277 128X
3. Abd-alrazzq, H.K., Ibrahim, M.S., Dawood, O.A.: Secure internet voting system based on public key kerberos. IJCSI Int. J. Comput. Sci. Issues **9**(2), 428–434 (2012)
4. Hazimeh, A., Mutaz,KH.K, Al-Talafha, K.M.: New Applied E-Voting System. **3** (2011, March 31)
5. Patel, T., Chokshi, M., Shah, N.: Smart device based election voting system endorsed through face recognition. **3**(11) (2017, November).ISSN: 2277 128X
6. Yadav, V.K., Batham. S., Mallik, A.K.: Kerberos based electronic voting System. ICNICT 2012, (2012, November)
7. Jefferson, Rubin, A.D., Simons, B., Wagner, D.: A security analysis of the secure electronic registration and voting experiment (SERVE). (2007, January)
8. Yang, C.-H., Tu, S.-Y. Yen, P.H.: Implementation of an electronic voting system with contactless IC Cards for small-scale voting. In: Fifth International Conference on Information assurance and Security, 978-0-7695-3744-3/09. ©IEEE
9. Niemoller, D.K.: Experience with voting machines in the Netherlands and Germany. Appendix 2K to the first report of Ireland's Commission on Electronic Voting (2004)
10. Nunamaker, J.A., Dennis, A.R., Valacich, J.S., Vogel, D.R., George, J.F.: Electronic meeting systems to support group work. **34**(7), 40–61 (1991, July). CACM
11. Nurmi, H., Salomaa, A., Santean: Secret ballot elections in computer networks. Comput. Sec. **10**(6), 553–560 (1991)
12. Khasawneh, M., Malkawi, M., Al-Jarrah, U., Hayajneh, T.S., Ebaid, M.S.: A biometric-secure e-voting system for election processes. In: Mechatronics and Its Applications 2008, ISMA 2008, 5th International Symposium, pp. 1–8. IEEE (2008, May 27–29)
13. Ondrisek, B.: E-voting security optimization. In: 42nd Hawaii International Conference on System Sciences (2009)
14. Pan, H., Hou, E., Ansari, N.: Ensuring voters and candidates' confidentiality in e-voting systems. IEEE. 978-1-61284-680-4/11 (2011)
15. Prasad, H.K., Halderman, J.A., Gonggrijp, R.: Security analysis of India's electronic voting machines. In: 17th ACM Conference on Computer and Communications Security CCS '10 (2010, Oct)

Minimax (Maximin) with Special Approach of Gamification in Higher Education

Showkat Nazir Lone, Babita Pandey and Aditya Khamparia

Abstract Gamification is the technique by which user can understood and solve complex problems easily. The main objective of the gamification is to enhance the learning capability of user; this is done by providing the topics in the form of game elements rather than providing as tutorials. It focuses on the play so as to give choice and information to learner for encouraging. It deals with the process of engaging the user with some sort of game with the motive of understanding some task or a problem. When the user is engaged more and more with it, the more the user will mastered the contents. This paper proposes the method of gamification of minimax (maximin) problem in higher education to help students to move in the depth of the minimax (maximin) problem. Minimax (maximin) gamification provides a clear and concise environment so that the learner would completely understand the minimax (maximin) problem while playing the game.

Keywords Collaborative · Competitive · Reward · Badges · Pretest · Posttest

1 Introduction

Gamification is the process of using game elements in non-gaming contents. The word "gamification" itself depicts that it has some relation with the games. Gamification does not mean to develop a complete game but to use some elements of the game for improving the experience of the user. In schools already there are

S.N. Lone · B. Pandey (✉)
Department of Computer Applications, Lovely Professional University, Phagwara, India
e-mail: shukla_babita@yahoo.co.in

S.N. Lone
e-mail: starshowkat@hotmail.com

A. Khamparia
Department of Computer Science and Engineering, Lovely Professional University, Phagwara, India
e-mail: aditya.khamparia88@gmail.com

A.K. Saini et al. (eds.), *ICT Based Innovations*, Advances in Intelligent Systems and Computing 653, https://doi.org/10.1007/978-981-10-6602-3_2

several game-like elements such as for correctly completing assignments students get reward in the form of points and then the points form “badges”, which are known commonly as grades. For desired behavior, students get rewarded and also for undesirable behavior students are punished using commonly system of reward.

From past few years, researchers had found that gamification plays a key role in teaching and learning process, complex problems can be easily taught to learners with the help of gamification. Gamification proves very beneficial in marketing as well as in higher education departments. In higher education, many courses or subjects can be taught using gamification, for example, data structure is an important subject in IT courses which can be taught using gamification. A variety of courses in computer science like data structure, programming, etc. are taught by using gamification (Lee 2011; Madani and Hooshyar 2014; Chang et al. 2011; Smith 1999). Gamification plays an important role in motivating the students for self-learning.

In this paper, we deployed the gamification technology for solving minimax (maximin) problem to easily and better understand the minimax (maximin) problem in a fun-loving manner.

The rest of the paper has been divided into following sections. Section 2 covers the literature review on learning techniques used by researchers. Section 3 deals with implementation of minimax (maximin) gamification. Results and observation findings are discussed in Sect. 4. Section 5 deals with the conclusion.

2 Literature Review

Over the last years, different forms of gamification have been applied to a variety of area such as education, energy management, multi-reservoir system, nonprofit marketing, well-being measures, security, online community, sports, event-driven surveillance, emoticon bin, online community, public transportation, banking, etc. The main goal of gamification is not to turn the systems into fully fledged games, but rather to apply some game elements in order to make the systems more motivating and engaging, or to alter users’ behavior in some meaningful way (Hakulinen et al. 2013). Gamification is a broad concept and uses methods such as achievement badges, leaderboards, points, levels, feedbacks, progress bar, Altruism, status, and power-ups. Gamification is a widely used method in education for engaging and motivating students. Educational system follows various learning theories and styles such as collaborative learning, constructivism learning, connectivism learning, and scaffolding Thomas and Berkling (2013). A detailed summary and comparative view of various areas where gamification is deployed for enhancing the performance is given in Table 1.

Table 1 Gamified system in various areas

Author name	Learning type	Game dynamics/mechanism
Stott et al. (2014)	Competitive and collaborative learning	Points, Rank, levels
Jayasinghe and Dharmaratne (2013)	Learning with Bloom's taxonomy	Rewards with gifts
Thomas and Berkling (2013)	Collaborative, Constructivism, Scaffolding, Connectivism, Critical Grasha-Riechmann Student Learning Style Scales, Kolb's Learning Style Inventory	Rewarding with Badges and leaderboards
Santos et al. (2013)	Collaborative learning	Rewards with Badges
Lee and Doh (2012)	Scaffolding learning	Feedback
Banfield et al. (2014)	Competitive learning	Rewards or reinforcements (e.g., extra credit or bonus points)
Vieira et al. (2012)	Public Transportation/Collaborative learning	Rewards with feedback
Blagov et al. (2013)	Collaborative environment	Rewards with points and prize
Silva (2010)	Education-e-learning-social gaming mechanics	Competitive learning Achievements
Lee et al. (2011)	Education	Mental, public and sensitive needs. Pitfalls
Kumar et al. (2012)	Education	Competitive learning feedback
Dominguez et al. (2013)	E-learning	Evaluation and Scores.
Laskowski et al. (2014)	Education	Rewards, achievements
Hakulinen et al. (2013)	Education—data structure	Competitive learning Rewards with Badges
Lee and Doh (2012)	E-learning	Scaffolding learning Feedback
Banfield et al. (2014)	Increasing student intrinsic motivation and self-efficacy	Competitive learning rewards or reinforcements (e.g., extra credit or bonus points)
Fujikawa and Min (2013)	For algorithm	Competitive and collaborative learning Rewards with badges
Su et al. (2013)	A mobile game learning system	Demographic variables, Learning achievements
Aplak et al. (2013)	Decisional process of energy management	Decision making and Evaluation process. Optimal strategies in governing energy policy
Madani et al. (2014)	Multi-reservoir system	Cooperative game theory challenges
Fredumann et al. [1]	Nonprofit marketing	Rewards with incentive
Hall et al. (2012)	Well-being measures	Points
Bista et al. (2013)	Online community	points and badges
Depura and Garg (2012)	Work/training	Leaderboards
Herzig and Ameling (2012)	Work/training	Leaderboards

(continued)

Table 1 (continued)

Author name	Learning type	Game dynamics/mechanism
Aud (2013)	Event-driven surveillance	Rewards with points
Berengueres et al. (2013)	emoticon bin, a recycle bin	Rewards with smiles and sounds
Vieira et al. (2012)	Public Transportation	Collaborative riding Rewards with feedback
Blagov et al. (2013)	corporate Web 2.0 systems in organizations	Collaborative environment Rewards with points and prize
Zhang et al. (2013)	Security	Competitive environment leader board

3 Design and Implementation of Minimax (Maximin) Gamification

The development of gamification of minimax (maximin) was done by using the Android Studio software which requires XML and Java coding for building applications.

3.1 Components of Gamification for Minimax (Maximin) Problem

In this section, we described various components of gamification of minimax (maximin) problem such as motivation and rewards, whereas motivation is further divided into competence, autonomy, and purpose as shown in Fig. 1.

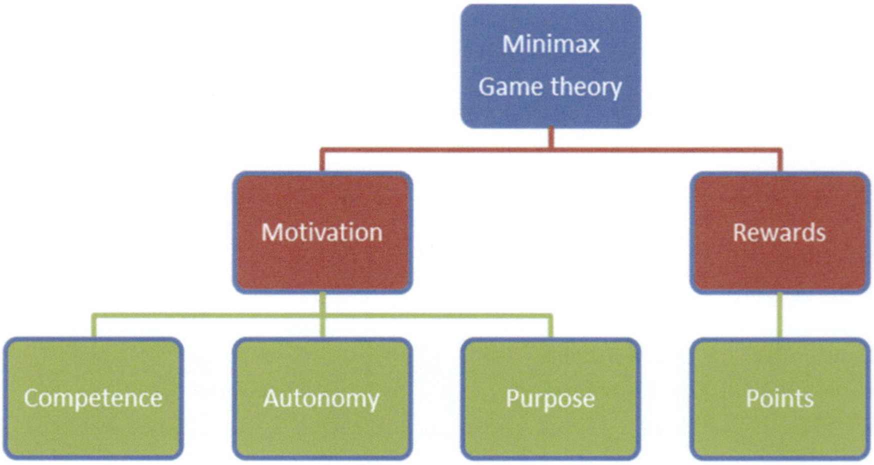

Fig. 1 Components of gamification

3.1.1 Motivation

By motivation, players take part in the game with their will and new tasks are given to them.

1 *Competence*: It deals with the process of engaging the user with some task related to the minimax (maximin) problem. When the user is engaged more and more with it, the more the user will be mastered the contents.
2 *Autonomy*: The user is given full freedom over the actions, which means that the user can take any action and user is not dependent for learning on anyone. The user gains experience and master the content by using the gamification.
3 *Purpose*: The main purpose of the gamification of minimax (maximin) problem is to learn the minimax (maximin) problem. By motivating the user with the gamification, user will learn the minimax (maximin) problem by engaging himself with gamification.

3.1.2 Rewards

After completion of each task, each player gets a reward for his achievements in the game.

Points: We motivate the user by applying the meaning gamification. We give some points as a reward so that to motivate and engage the user.

3.1.3 Preview of Minimax (Maximin) Gamification

The screen shown in Fig. 2 is the first page of the application, and the functions of the given buttons are given as follows:

Play: User can start playing the minimax (maximin) game by clicking on the play button.

Help: *U*ser can take help regarding gameplay by clicking on the help button.

About: This button contains the information about the developer of the application.

The screenshot shown in Fig. 3 is the main screen of the application for playing the minimax (maximin) game, it also provides clear and concise hints to the user so that he would completely understand the minimax (maximin) problem while playing the game.

4 Results

The method of implementing gamification of minimax in management learning platform is shown in Fig. 4.

Fig. 2 First page of the application

Fig. 3 Main screen for playing the minimax (maximin) game

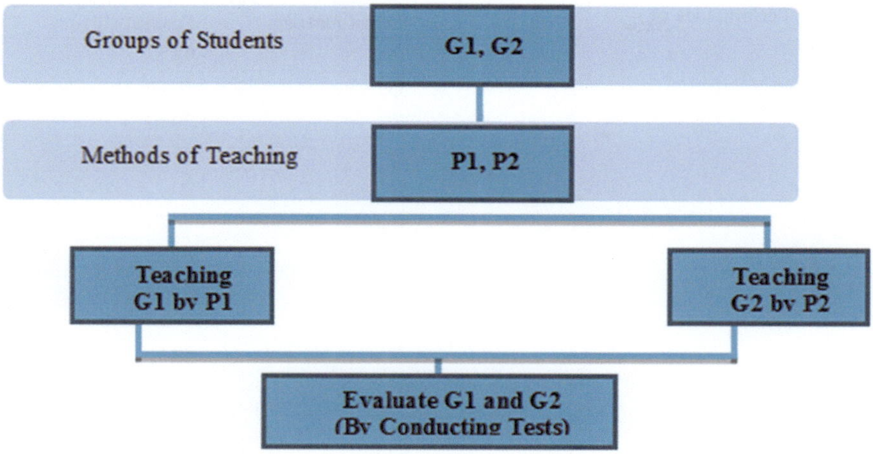

Fig. 4 Method of implementing gamification of minimax

A study was conducted upon 30 students of final year of Bachelor's in Business Administration, which were divided into two groups: an experimental and control groups based on homogeneity test. Each group had 15 students. The experimental group (G1) was gamified, while the other group (G2) followed the traditional method of teaching, i.e., whiteboard teaching.

The pretest and posttest consist of 30 questions based on minimax problem. The performance of the students in pre- and posttest was compared and calculated by using t test. The performances of students in pretest and posttest of G1 (gamified) are shown in second and third columns of Table 2, respectively.

Table 2 Performance of group 1

S. no.	Pretest (%)	Posttest (%)
1	56.67	70
2	63.33	86
3	41	56.67
4	61	83.33
5	73.33	90
6	53.33	83.33
7	21	50
8	13	36.67
9	24	50
10	43.33	63.33
11	59.67	93.33
12	63	70
13	26.67	46.67
14	67	93.33
15	81	90

Table 3 Performance of group 2

S. no.	Pretest (%)	Posttest (%)
1	76	93.33
2	55	70.67
3	30	20.67
4	45	58
5	51	90
6	46.67	50.67
7	26.67	30.33
8	24	30.67
9	25	50.67
10	56.67	59
11	69	71
12	57	80.33
13	64.67	50.67
14	58	60.33
15	64	74.33

The performances of students in pretest and posttest of G2 (teaching using presentations) are shown in second and third columns of Table 3, respectively.

The t tests are applied at four stages; the results are shown in Table 3.

(1) t test is applied on the pretest results of group 1 ($M = 14.94$, SD = 5.61) and group 2 ($M = 14.97$, SD = 5.9); for this, the $t = 0.0147$ and $p = 0.9884$ are obtained which show that the groups 1 and 2 are not significantly different.
(2) t test is applied on the posttest results of group 1 ($M = 25.34$, SD = 4.86) and group 2 ($M = 20.49$, SD = 7.48); for this, the $t = 2.1058$ and $p = 0.044$ are obtained which show that the groups 1 and 2 are significantly different.
(3) t test is applied on the pre- and posttest results of group 1 (pretest result: $M = 14.94$, SD = 5.61) and group 1 (posttest result: $M = 25.34$, SD = 4.86); for this, the $t = 5.42$ and $p = 0.0001$ are obtained which show that the results are extremely significant.
(4) t test is applied on the pre- and posttest results of group 2 (pretest result: $M = 14.97$, SD = 5.9) and group 2 (posttest result: $M = 20.49$, SD = 7.48); for this, the $t = 2.1058$ and $p = 0.0298$ are obtained which show that the results are statistically significantly (Table 4).

Table 4 Comparisons of results of groups 1 and 2

Test	Group 1		Group 2		T and p values
	Mean (M)	SD	Mean (M)	SD	
Pretest scores (0 and 30)	14.94	5.61	14.97	5.59	$t = 0.0147$; $p = 0.9884$ (not significant)
Posttest scores (0 and 30)	25.34	4.86	20.49	7.48	$t = 2.1058$; $p = 0.044$ (significant)
t and p value	$t = 5.42$, p = 0.0001 extremely significant		$t = 2.28$, $p = 0.0298$ statistically significant		

The posttest results of G1 and G2 based on showing the individual percentages acquired by each student are shown in Fig. 5 and Table 5.

On the basis of performance evaluation, we can say that proposed method that is gamification of minimax (maximin) is best for enhancing the learning capability of the user (Fig. 6).

4.1 Performance Comparison

Although many applications in computer science (CS) are available, gamification is implemented for enhancing learning of students. All these existing systems are for CS domain, run on computer, complex in nature, and not measure the performance of student in terms of time, whereas the minimax system is for management domain, mobile application, easy to understand and measure the performance of students in time also. A comparison of the minimax with other existing gamified system is given in Table 6.

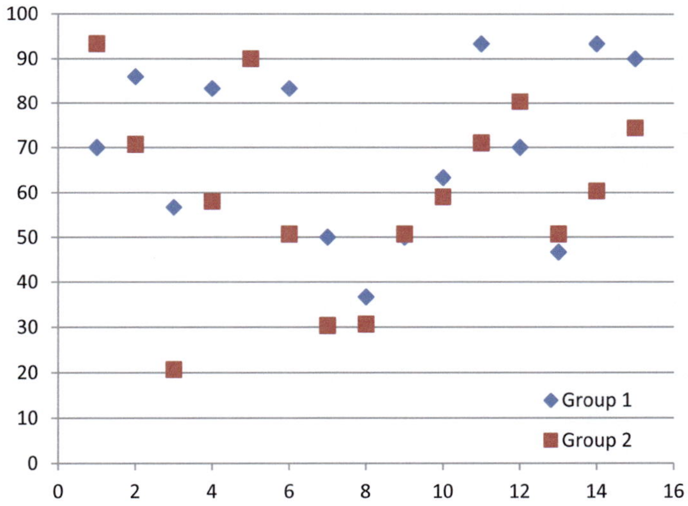

Fig. 5 Performance of students in posttest of groups G1 and G2

Table 5 Performance of students in posttest of groups G1 and G2

Percentage score	Group 1	Group 2
0–20	0	0
21–40	1	3
41–60	4	5
61–80	3	4
81–100	7	3

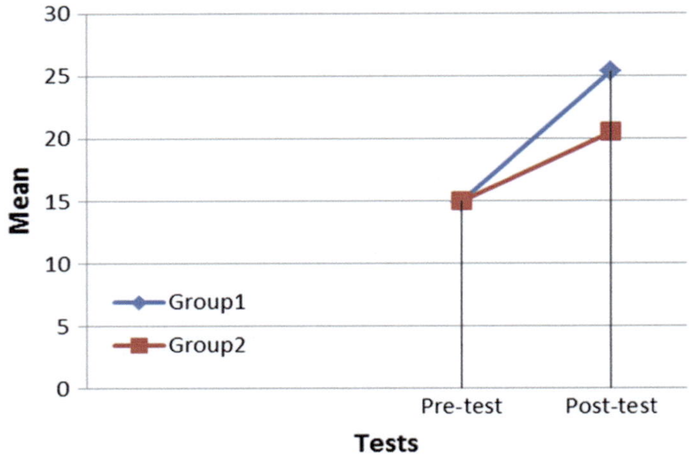

Fig. 6 Pre–Post-analysis

Table 6 Performance comparison of minimax game

Author name	Domain	Platform	Understandability	Performance measure factors
Silva (2010)	CS	Computer	Complex	NA
Lee et al. (2011)	CS	Computer	Complex	NA
Kumar et al. (2012)	CS	Computer	Complex	NA
Dominguez et al. (2013)	CS	Computer	Complex	NA
Laskowski et al. (2014)	CS	Computer	Complex	NA
Stott et al. (2014)	CS	Computer	Complex	NA
Jayasinghe and Dharmaratne (2013)	CS	Computer	Complex	NA
Thomas and Berkling (2013)	CS	Computer	Complex	NA
Santos et al. (2013)	CS	Computer	Complex	NA
Hakulinen et al. (2013)	CS	Computer	Complex	NA
Lee and Doh (2012)	CS	Computer	Complex	NA
Tabata and Hashimoto (2011)	CS	Computer	Complex	NA
Banfield et al. (2014)	CS	Computer	Complex	NA
Fujikawa and Min (2013)	CS	Computer	Complex	NA
Su et al. (2013)	CS	Computer	Complex	NA
Our approach (2015)	Management domain	Mobile application	Easy	Score and time

5 Conclusions

On the basis of performance evaluation, we can say that proposed method that is gamification of minimax (maximin) proves successful in achieving its objectives. Minimax (maximin) gamification provides a clear and concise environment for the learner to completely understand the minimax (maximin) problem while playing the game. The minimax (maximin) gamification enhanced the learning capability of users more as compared to other modes of teaching.

One needs to identify the purpose, scope, and domain of the gamification as well as a source of the domain knowledge using a systematic approach. The evolution of gamification and computer-based educational systems should support the import and export of awareness in a standard format with common semantics.

References

1. Freudmann, E.A., Bakamitsos, Y.: The role of Gamification in non-profit marketing: an information processing account. Procedia-Soc. Beh. Sci. **148**, 567–572 (2014)
2. Lee, J.J., Hammer, J.: Gamification in education: what, how, why bother. Acad. Exch. Q **15** (2), 1–5 (2011)
3. Madani, K., Hooshyar, M.: A game theory–reinforcement learning (GT–RL) method to develop optimal operation policies for multi-operator reservoir systems. J. Hydrol. **519**, 732–742 (2014)
4. Chang, W.C., et al.: Learning Kruskal's Algorithm, Prim's Algorithm and Dijkstra's Algorithm by board game. In: Proceeding of International Conference on Web-Based Learning, pp. 231–239 (2011)
5. M. K. Smith.: 'Learning theory', the Encyclopedia of informal education. Available at: http://www.infed.org/biblio/b-learn.htm (1999). Last Accessed 25 June 2010
6. Silva, E.: Gamifying learning with social gaming mechanics. Masie Learn. Center Perspect. pp. 212–218 (2010)
7. Simoes, J., et al.: A social gamification framework for a K-6 learning platform. Comput. Hum. Behav. **29**, 345–353 (2013)
8. Ang, C.S., et al.: Linking pedagogical theory of computer games to their usability. Int. J. E-Learn. **7**, 533–558 (2008)
9. Aplak, H.S., Sogut, M.Z.: Game theory approach in decisional process of energy management for industrial sector. Energy Convers. Manag. **74**, 70–80 (2013)
10. Laskowski, M., Badurowicz, M.: Gamification in higher education: a case study, human capital without borders: Management, knowledge and learning for quality of life knowledge and learning 25–27 June 2014. Portoroz, Slovenia International Conference pp. 971–974 (2014)
11. William, D.: Formative Assessment: Getting the Focus Right. Edu. Assess. **11**(3/4), 283–289 (2006)
12. Browne, K., et al.: Gamification and serious game approaches for adult literacy tablet software. Enter. Comput. **5**, 135–146 (2014)
13. Kumar, B., Khurana, P.: Gamification in education-learn computer programming with fun. Int. J. Comput. Dis. Syst. **2**(1), 46–53 (2012)
14. Fritsch, R., Fritsch, G.: The four color theorem, history, topological foundations and idea of proof. pp. 1–260. Springer-Verlag, New York (1998)
15. Gasland, M.: Game mechanic based e-learning. Science and Technology. Master Thesis, pp. 75–81 (2011)

16. Banfield, J., Wilkerson, B.: Increasing student intrinsic motivation and self-efficacy through Gamification Pedagogy. The Clute Institute International Academic Conference Orlando, Florida, USA, pp. 138–145, (2014)
17. Dominguez, A.: Gamifying learning experiences: practical implications and outcomes. Comput. Edu. **63**, 380–392 (2013)
18. Stott, A. Neustaedter, C.: Analysis of Gamification in education, school of interactive arts and technology, Simon Fraser University, 250–1345010 (2014)
19. Su, C.H., Cheng, C.H.: A mobile game-based insect learning system for improving the learning achievements. Procedia—Soc. Beh. Sci. **103**, 42–50 (2013)
20. Bellotti, F., et al.: Serious games and the development of an entrepreneurial mindset in higher education engineering students. Ent. Comput. pp. 1–10 (2014)
21. Berengueres, J., Alsuwairi, F., Zaki, N., Ng, T.: Gamification of a Recycle Bin with Emotions. IEEE. pp. 83–84 (2013)
22. Hakulinen, L., Auvinen, T., Korhonen, A.: Empirical study on the effect of achievement badges in TRAKLA 2 Online Learning Environment IEEE. pp. 47–54 (2013)
23. Hall, M., O.kimburgh, S., Haas, C., Caton, S.: Towards the Gamification of Well Herzig. In: P.,Ameling, M. A Generic Platform for Enterprise Gamificatio,. Joint working Conference on Software Architecture and 6th European Conference on Software Architecture. pp.219–223 (2012)
24. Jayasinghe, U., Dharamratne, A.: Game Based Learning VS Gamification from the higher education students perspective. IEEE Int. Conf. Teach. (2013)
25. Lee, H., Yim Doh, Y.: A study on the relationship between educational Achievement and emotional engagement in a gameful interface for video lecture Systems. In: International Symposium on Ubiquitous Virtual Reality, pp. 34–37 (2012)
26. Thomas, C., Berling, K.: Redesign of a Gamified software engineering course. In: International Conference on Interactive Collaborative Learning (ICL), pp. 778–786 (2013)
27. Depura, K., Garg, M.: Application of online Gamification to new hire onboarding. Third International Conference on services and Emerging markets. pp. 153–156 (2012)
28. Vieira, V., Fialho, A., Martinez, V., Brito, J., Duran, L.: An Exploratory study on the use of collaborative riding based on Gamification as a support to Public Transportation. In: Brazilian Symposium on Collaborative Systems, pp. 84–92, (2012)
29. Zhang, K., Dong, S., Zhu, G., Corporan, D., Mcmullan, T., Barrera, S.: Pico CTF 2013 Toaster Wars. Int. Games Inn. Conf. (IGIC). pp. 293–299 (2013)
30. Aud, J.: Lean tail labs, Austin Texas: Gamification—a real world, example, IEEE conference on gaming. pp. 128–134 (2013)
31. Bista, S.K., Nepal, S., Paris, C.: Data abstraction and visualisation in *next step*: experiences from a government services delivery trial. IEEE Int. Congress Big Data 267–270 (2012)
32. Srikanth, A.: Making banking fun—customer experience management. TCS BaNCS Marketing, Communications and Research (2012)
33. Tabata, H., Hashimoto, H.: WBL for technical skill of MONODZUKURI, SICE annual conference. September 13–18, Waseda University, Tokyo, Japan (2012)
34. Pandey B., Mishra, R.B., Khamparia, A.: CBR based approach for adaptive learning in E-learning system. In: IEEE conference on Computer science and engineering (Asia pacific world conference, Fiji), doi:10.1109/APWCCSE.2014.7053877, pp. 1–6 (2014)
35. Khamparia A., Pandey B., Pardesi V.: Performance analysis on agricultural ontology using SPARQL query system. In: IEEE conference on data mining and intelligent computing (ICDMIC), doi:10.1109/ICDMIC.2014.6954258, pp. 1–5 (2014)
36. Khamparia A., Pandey B.: Adaptive e-learning based Udutu system for object oriented programming. In: International Journal of Applied Engineering Research, **10**(2) pp. 1439–1442 (2015). ISSN 0973-4562
37. Pandey B., Khamparia A., Raman.: Performance analysis of adaptive intelligent tutoring system. In: International Journal of Applied Engineering Research, **10**(63) (2015). ISSN 0973-4562

Comparing the Behavior of Oversampling and Undersampling Approach of Class Imbalance Learning by Combining Class Imbalance Problem with Noise

Prabhjot Kaur and Anjana Gosain

Abstract Class imbalance learning is a recent topic, which helps us to detect the classes from unbalanced datasets. In various real scenarios, where we need to find the exceptional cases like credit card problem, brain tumor detection, etc., the traditional classification algorithms fail because they are designed in such a way that either their results are overwhelmed by the bigger class and or they ignore the smaller class as a noise and avoid it. In recent studies, it has been found that class imbalance problem itself is not a problem but there are certain other data distribution complexities, which when combined with the class imbalance problem degrade the performance of classifier. One of the major issues is noise in the data, which is a part of every real data in one form or another. This paper compares the oversampling and undersampling approaches of class imbalance learning in noisy environment and tries to find out which is the better approach in such case.

Keywords Class imbalance problem · Class imbalance learning · Undersampling · Oversampling · RUS · SMOTE

1 Introduction

Classification is a tool to classify datasets into various classes, and it worked nicely if it is applied on a balanced data, i.e., data-set with almost same size of the classes but in the case of unbalanced datasets, wherein there is a huge difference between the sizes of classes, these algorithms fail to detect the classes. The reason behind the failure is that these algorithms are designed in a way to deal with balanced classes so when they are applied on the dataset with unequal classes, their results always

P. Kaur (✉)
Maharaja Surajmal Institute of Technology, Janakpuri, New Delhi, India
e-mail: thisisprabhjot@gmail.com

A. Gosain
USICT, Guru Gobind Singh Indraprastha University, New Delhi, India
e-mail: anjana_gosain@hotmail.com

A.K. Saini et al. (eds.), *ICT Based Innovations*, Advances in Intelligent Systems and Computing 653, https://doi.org/10.1007/978-981-10-6602-3_3

deviate toward the bigger class or due to the design of traditional classification algorithms they ignore the smaller class as noise. This classification problem is called class imbalance problem (CIP) and have been identified a decade ago. This problem can be risky in case of many real-time problems if we use traditional classification algorithms to solve them like in case of the detection of medical diagnosis (Brain Tumor detection), to detect fraudulent telephone class, credit card transactions, etc. or in any case where there is a need to detect the exceptional cases [1, 2]. To tackle the class imbalance problem, researchers have taken various ways and a new field of research is identified under the name class imbalance learning, which is a recent field of research and is evolving day by day.

Before starting with the detailing of class imbalance learning (CIL), we want to share some terminology of this field referred in various papers. The smaller class is known as minority class and bigger class is known as majority class. CIP actually exists because the problem is to identify minority class. The ratio of majority class instances to minority class instances is called imbalance ratio. The problem becomes more risky as this ratio increases, because dataset is highly unbalanced in such situation. Various approaches taken by researchers to solve the class imbalance problem are data-level approach (Preprocessing techniques), algorithm-level approach, and their hybrid forms [3, 4] (as shown in Fig. 1). In data-level approach, the researchers have tried to balance the datasets before applying traditional classification algorithms so that results may not be impacted by the majority class. In algorithm-level approach, the researchers have worked on the internal algorithm structure and tried to remove the sensitivity of algorithm toward the majority class so that results of classification algorithms should not deviate toward

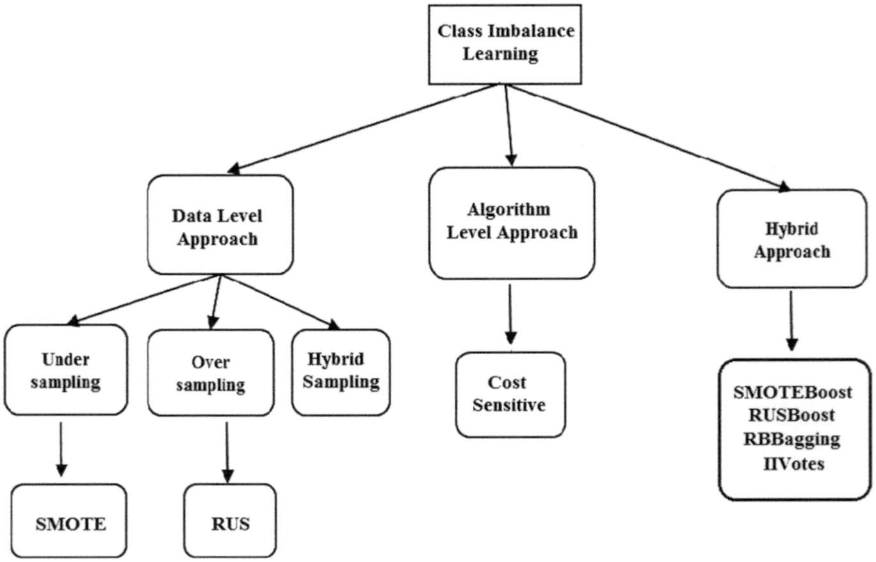

Fig. 1 Classification of class imbalance learning approaches

the majority class. These algorithms come under the category of cost-sensitive algorithms [5–10]. Third approach is the hybrid form, which is the combination of data-level and algorithm-level approaches [11–14]. In the recent research [15–19], it has been found that CIP itself is not a problem but other data distribution complexities like overlapping classes and noise also degrade the performance of classifiers designed for CIP. Real datasets always contain noise in one form or another, e.g., in the form of missing, changed, or vague values in the datasets like in case of medical images because of the acquisition of images through electronic machines, there is a problem of vague values. In this paper, we are comparing two very famous data-level approaches, i.e., oversampling and undersampling in the noisy environment to find out which one is the better approach if unbalanced data distribution also contains noise. Section 2 explores the data-level techniques in detail and Sect. 3 compares and discusses the results after implementing oversampling and undersampling techniques in noisy environment followed by conclusion in Sect. 4.

2 Data-Level Approaches

Data-level approaches [20] are divided into three sections. Oversampling, undersampling, and hybrid sampling (Fig. 1). Oversampling is the process of increasing the number of instances into the minority class either randomly through replication of the same data or generating synthetically by using some technique to improve the imbalance ratio so that same classification algorithms can be used to classify the data. The advantage of this technique is that there is no loss of any important information from the dataset and the original dataset is retained although new information is added to it to balance the data. The limitation of oversampling is that it takes more time to execute as compared to undersampling approach as we are increasing the number of instances. It may also cause the problem of overfitting in case it replicates the same instances. Another issue with oversampling is that as our target in case of CIP is to identify minority class, oversampling changes the minority class by adding more instances and actually changing the class that we want to identify, which may not be acceptable in some critical real-time problems.

In the undersampling approach, the working area in the dataset is majority class wherein the instances from the majority class are removed either randomly or by using some technique to balance the classes and then traditional classification algorithms are applied to classify the data. Hybrid sampling is the combination of oversampling and undersampling approaches, wherein both the techniques, i.e., oversampling and undersampling, are applied on the unbalanced data to improve the imbalance ratio and then classes are identified using traditional classification algorithms. The main advantage of this technique is that it is fast and simple approach as compared to oversampling. But as the instances are removed from the majority class, it may lead to the loss of some potentially useful information contained in the majority class. To compare the oversampling and undersampling approaches in the noisy environment, the paper is using SMOTE (Synthetic

Minority Oversampling Technique) [21] and RUS (Random Undersampling) [11] algorithms.

2.1 Synthetic Minority Oversampling (SMOTE) [21]

SMOTE is an oversampling approach which synthetically generates instances by randomly selecting instances from the minority class and using interpolation method to generate instances between selected point and its nearby instances. In this process, every minority class instance is considered and new minority class instances are generated along the line segment joining its k-nearest neighbors. The number of synthetic instances is generated based on the percentage of oversampling required. The steps of the algorithm are as follows:

SMOTE Algorithm:

1. Load dataset and identify the minority and majority class;
2. Calculate the number of instances to be generated based on the percentage of oversampling;
3. Identify a random instance from minority class and find its nearest neighbors;
4. Select any of the one nearest neighbors and find the difference between random instance and its selected nearest neighbor;
5. Multiply the difference by a random number generated between 0 and 1;
6. Add this difference to the selected random instance;
7. Repeat the process from 3 to 6 till the number of instances is generated as per the given percentage.

2.2 Random Undersampling (RUS) [11]

RUS is a very simple undersampling approach which randomly removes instances from the majority class to balance the dataset before applying classification technique. The concept of RUS is very simple and it is fast as compared to SMOTE. The only limitation with this technique is that it can remove the important information contained in the majority class, which may not be acceptable in some cases. It can be illustrated in Fig. 2.

In the figure, green line shows the ideal decision boundary that we want to identify and blue line shows the actual results obtained. Left side of the figure shows the result of classification after applying a general machine learning algorithm without undersampling and right section shows the result after applying undersampling. By applying undersampling on the majority class, some informative majority class information is removed and it caused the blue decision boundary to be slanted, causing some majority class instances to be classified as minority class instances wrongly.

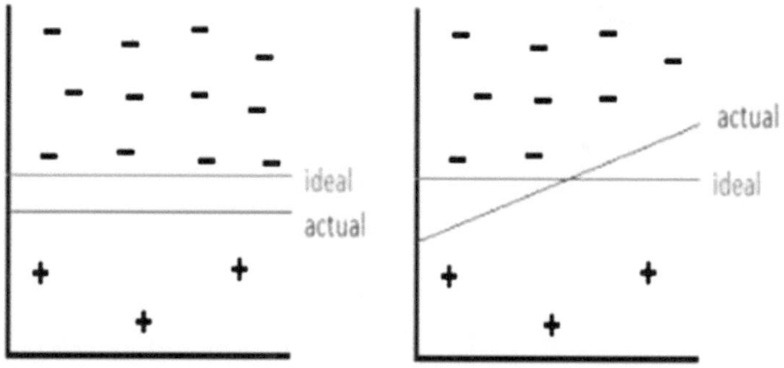

Fig. 2 Classification with and without using undersampling

RUS Algorithm:

1. *Load dataset and identify the minority and majority class;*
2. *Calculate the number of instances to be removed based on the percentage of undersampling;*
3. *Identify a random instance from majority class and remove it from the majority class;*
4. *Repeat step 3 till the number of instances removed as per the given percentage.*

3 Empirical Comparison Between Oversampling (OS) and Undersampling (US) Approach in Noisy Environment

To observe the behavior of these approaches for the combining effect of class imbalance and noise, the paper is considering SMOTE and RUS algorithms. Dataset used in the paper is synthetically generated by introducing 70% random noise as shown in Fig. 3. The tools used to do the experiment are MATLAB 2010b [22] and WEKA 3.7.12 [23]. MATLAB tool is used to do the preprocessing using oversampling and undersampling approaches and WEKA is used to do the classification using decision tree under stratified tenfold cross-validation mode. C4.5 is used as a classifier, which is an efficient classifier for balanced datasets. It is implemented in Weka with the name as j48 classifier. Imbalance ratio of the dataset is 1:12 without considering noise, which is highly unbalanced situation.

Results are recorded in Table 1 by varying the percentage of undersampling and oversampling with 10, 40, 50, and 80%. The performance parameter used to assess the algorithms is AUC (Area under the ROC curve) [93–96], which have been extensively used by the researchers in this field. The classification accuracy increases with the increase in AUC value.

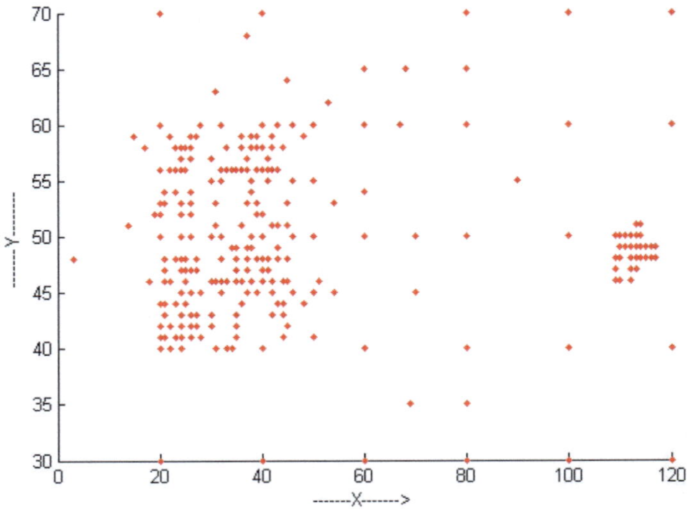

Fig. 3 Synthetic dataset used in the experiment

Table 1 Results of oversampling and undersampling approaches in noisy environment

Algorithm	%age of US		Imbalance ratio	AUC
RUS	10		14.6	0.945
	40		10.5	**0.923**
	50		9.2	**0.927**
	80		5.1	0.899
SMOTE	%age of OS	Noise points generated	Imbalance ratio	AUC
	10	2	13.13	0.898
	40	**6**	7.91	**0.938**
	50	**9**	7.21	**0.916**
	80	6	4.72	0.968

As per traditional classification, if we are decreasing the imbalance ratio then the classification accuracy increases. But in the current study with the noisy dataset, the algorithms are behaving in a completely different manner. In case of RUS, it is observed that as the level of undersampling is increased, which makes the classes more balanced, the performance of algorithm degrades (as shown by the AUC field of Table 1). The reason behind this is that RUS randomly removes instances from the majority class, and as the percentage of undersampling increases there is a possibility of removal of more informational data and the impact of noise points contains the majority class is more. As a result of this situation, the performance of the classifier degrades. In Table 1, it is observed that when the undersampling is changed from 10 to 40%, the performance of the algorithm degrades whereas from 40 to 50%, its performance enhances and again with 80% undersampling when the data is more balanced, its performance again degrades. Its behavior is completely unpredictable and completely depend upon the removal of type of instances whether the instances which are randomly selected are informational data or noise.

In case of SMOTE, the performance of classifier increases with the increase in the percentage of oversampling except the case when the percentage of oversampling is changed from 40 to 50%. The reason behind this transition is that the SMOTE technique generates a synthetic point between a random number and its nearest neighbor. As the minority class contains noise, so in the process of SMOTE, some noise points are also generated. The number of noise points generated at various levels is listed in Table 1. At 50% oversampling, the number of noise points generated is nine which is more than that is generated in the case of 40%. Although the classes are more balanced at 50% oversampling, the impact of noise has degraded the performance of the classifier.

If we compare the behavior of both the techniques in the noisy environment, the performance of SMOTE increases when the classes are changing from unbalanced mode to balance mode, whereas the RUS behaves in the reverse manner, i.e., its performance degrades when the classes become more balanced. From these observations, we can say that in the noisy environment there is a need to intelligently select the instances which are to be removed or generated in the datasets to make the classes balanced. As SMOTE and RUS, both these techniques are blindly generated and remove the instances from the dataset so if we compare the performance of these two, then SMOTE behaves better than RUS in the noisy environment so we can say that the behavior of oversampling technique is more robust than undersampling.

4 Conclusion and Future Work

In this paper, SMOTE (Oversampling) and RUS (Undersampling) data-level approaches are compared in the noisy environment. A synthetic dataset containing 70% random noise is used to compare the performance using the MATLAB 2010b and Weka 3.7.12 tool. It has been observed that oversampling approach (SMOTE) behaved in a robust manner than undersampling (RUS) approach in the noisy environment. In the future, we will analyze more intelligent undersampling and oversampling approaches for the combining effects of class imbalance with other data complexities like noise and class overlapping.

References

1. Yong, Y.: The Research of Imbalanced data-set of sample sampling method based on K-means cluster and Genetic algorithm. Energy Procedia, vol. 17, pp. 164–170. Sciverse ScienceDirect (2012)

2. Garcia, V., et. al.: The class imbalance problem in pattern classification and learning. Pattern analysis and learning group, Conreso Espanol de Informatica; pp. 283–291
3. Napierala, K., et. al.: Learning from Imbalance data in presence of Noisy and Borderline Examples. RSCTC, LNAI 6086, pp. 158–167, Springer-Verlag, Berlin Heidelberg (2010)
4. Satyashree, K.P.N.V., Murthy, J.V.R.: An exhaustiv literature review on class imbalance problem. Int. J. Emerging Trends Technol. Comput. Sci. **2**(3), 109–118 (2013)
5. Fernandez, A., et al.: A study of the behaviour of linguistic fuzzy rule base classification systems in the framework of imbalanced data-sets. Fuzzy Sets Syst. **159**(18), 2378–2398 (2008)
6. Fernandez, A., et al.: Hierarchical fuzzy rule base classification system with genetic rule selection for imbalanced data-sets. 50, 561–577 (2009)
7. Batuwita, R., Palade, V.: FSVM-CIL: fuzzy support vector machine for class imbalanced learning. IEEE Trans. Fuzzy Syst. **18**(3), 558–571 (2010)
8. Zhao, Z., Zhong, P., Zhao, Y.: Learning SVM with weighted maximun margin criterion for classification of imbalanced data. Math. Comput. Model. **54**, 1093–1099 (2011)
9. Galar, M., et al.: Dynamic classifier selection for one-vs-one strategy: avoiding non-competent classifiers. Pattern Recogn. **46**, 3412–3424 (2013)
10. Gu, X., et al.: New fuzzy support vector machine for the class imbalance problem in medical data-sets classification. The Scientific World Journal, vol. 2014, pp. 1–12, Hindawi Publishing Corporation (2014)
11. Seiffert, C., et al.: RUSBoost: a hybrid approach to alleviating class imbalance. IEEE Trans. On Sys. Man and Cyber.-Part A **40**(1), 185–197 (2010)
12. Hido, S., Kashima, H., Takahashi, Y.: Roughly balanced bagging for imbalanced data. Stat Anal Data Min **2**, 412–426 (2009)
13. Blaszczynski, J., Deckert, M., Stefanowski, J., Wilk, S.: Integrating Selective pre-processing of imbalanced data with ivotes ensemble. Rough sets and Current trends in Computing (Lecture notes in Computer Science Series 6086), Springer-Verlag, pp. 148–157, (2010)
14. Chawla, N., V., Lazarevic, A., Hall, L., O., Bowyer, K., W.: SMOTBoost: improving pridiction of the minority class in boosting. In proc. Knowledge Discovery databases, pp. 107–119 (2003)
15. Saez, J.A., Luengo, J., Stefanowksi, J., Herrera, F.: Managing BorderLine and Noisy examples in Imbalanced Classification by combining SMOTE with Ensemble filtering. IDEAL 2014, LNCS 8669, pp. 61–68, Springer, (2014)
16. Garcia, V., et. al.: Combined effects of Class Imbalance and Class Overlap on Instance-based Classification, IDEAL 2006, LNCS, vol. 4224, pp. 371–378, Springer Heidelberg (2006)
17. Prati, R.C., Batista, G.E., Monard, M.C.: Class imbalance versus class overlapping: an analysis of a learning system behaviour, In: Proc. 3rd Mexican International Conference on Artificial Intelligence, pp. 312–321, (2004)
18. Japkowicz, N.: Class imbalance: are we focussing on the right issue?, In proc. International workshop on learning from imbalanced data-sets II, (2003)
19. Jo, T., Japkowicz, N.: Class imbalance versus small disjuncts. SIGKDD Explorations **6**, 40–49 (2004)
20. Batista, G.E.A.P.A., et al.: A study of the behaviour of several methods for balancing machine learning training data. SIGKDD Expl. Newl. **6**(1), 20–29 (2004)
21. Chawla, N.V., et al.: SMOTE: synthetic minority over sampling technique. J. Artif. Intell. Res. **16**, 321–357 (2002)
22. MATLAB version 7.10.0. Natick, Massachusetts: The MathWorks Inc., 2010
23. Mark, H., Eibe, F., Geoffrey, H., Bernhard, P., Peter, R., Ian, H. Witten: the weka data mining software: an update. SIGKDD Explorations **11**(1) (2009)
24. Wang, Q.: A hybrid sampling SVM approach to imbalanced data classification. Abstract and Applied Analysis, vol. 2014, pp 1–7, Hindwani Publishing Corporation, (2014)

Proposed ICT-Based Transportation Model: EEG

Archana Singh, Sakshi Goel, Hina Gupta and Vikas Deep

Abstract Our cities are growing at an enormous rate, industry setup, and roads to link different parts of cities. The expansion of cities led to many problems primarily among them is the problem of traffic congestion which has adverse effect on social, environmental, and economical sustainability. In this paper, we have proposed a new model called as EEG, i.e., electronic roads, electronic wallets, and green roads. The motive behind our research is to apprehend the traffic system by making our roads intelligent. This can be done with the help of solar roadways where instead of implementing solar panels on the roofs of the building, we are using them on concrete roads. Another big thing that we are proposing is how to avoid congestion on toll booths by making use of smart wallet in order to save time.

Keywords ICT · Transport system · E-roads · E-wallet · Green roads

1 Introduction

ICT (information and communications technology—or technologies) technology is the integration of interacting device or application as well as the various services and applications associated with them. It facilitated the use of information through telecommunications and other mediums. This technology is springing up like mushroom in all the fields like the Internet, wireless networks, cell phones, educational fields, fashion, transportation, and agriculture. The implementation of ICT technology is still challenging in various developing and developed countries. The impact of the usage of internet worldwide is quite reflective. Information technology is a well-known term for the updated study of procedures, resources, and methods for transmission, managing, processing, saving, and presentation of data and information. Information Technology (IT) is a term that accentuates the components (hardware equipment) and programs (software) which allow us to access,

---

A. Singh (✉) · S. Goel · H. Gupta · V. Deep
Amity University Uttar Pradesh, Noida, India
e-mail: archana.elina@gmail.com

A.K. Saini et al. (eds.), *ICT Based Innovations*, Advances in Intelligent Systems and Computing 653, https://doi.org/10.1007/978-981-10-6602-3_4

Table 1 Use of ICT in different fields

Use of ICT in different fields	Impact on transportation
Use of e-banking, e-shopping, e-entertainment, e-education, and e-government services	Usage of such services reduces the demand for routine travel and transactions, but demand travel to central places may increase. For example, travel to shopping complexes for entertainment and shopping
E-networking solutions including socialization of public relationships, chatting via Internet, etc.	Usage of such solution reduces the travel among personal communities but may increase the demand for social communities established through social network
Work from home	Usage of this service reduces the travel of people working at different places and reduces road congestion
Online working or tele-center	Usage of this service reduces the travel of people to go to different centers for work
E-meeting (video conferencing)	Usage of this service reduces the need for travel to a limited extent but in some cases, this travel cannot be avoided

fetch, formulate, manipulate, and display information electronically. The Communication Technology (CT) is a term, which highlights telecommunication equipments with the support of which we can send, receive, search, and access the information. Collectively, it termed as the information and communication system. Table 1 shows the role of ICT in different fields.

1.1 Need of ICT in Traffic

Transport and traffic are the two domains whose properties are linked with spatial regions. This makes their maintenance and implementation difficult with the absence of adequate infrastructure and database GIS (geographical information system) (Durokovic 2011). There have been certain global concerns over the years that have blossomed. One of the major concerns is traffic control demand which is growing steadily. Another problem is that time spent on the road and in traveling is also unbearable. Third, the impact of traffic on environment, i.e., air and noise pollutions, caused due to vehicles. Many aspects of society were improvised by the implementation of the progressive technology, but there are still many disadvantages to it. A major barrier that exists in the progress of our country is the traffic and transport systems. Therefore, the improvement in transport system would be the focused on three main themes.

(a) **Traffic congestion**: A bottleneck on many international routes, congestion in cities, is a prime area of concern and prompt action needed to be taken.

(b) **Pollution and Health**: The pollution has now become a global concern. All countries are taking quantifiable measures to overcome this problem; otherwise, it would heavily or deadly impact our future generation. If the green environment got destroyed, then it would lead to the extinction of mankind. According to report from WHO (World Health Organization), there is a hole in our ozone layer over Antarctica that was due to greenhouse effect of gases, and the icebergs in Antarctica are melting at high rate. With this alarming situation, there would be a great possibility of floods and high tides over the sea shore regions.

(c) **Safety**: The safety is another obstacle in the management of transport system. There is a dire need that people should opt for public transportation as much as possible. Increases in traffic volumes, rehabilitation, and inevitable road maintenance are the major problems that seem to be very challenging.

Using intelligent transportation system in the industrial area should be improvised with the progression of routes; facilitate its use, with the introduction of flexible office timings so that at the same time many vehicles won't be on road this would help in the reduction of the cost and the most important would prevent accidents. The extant research suggests that ICT reduces congestion or traffic on roads which makes the travel hectic and boredom [1]. In 2002, a survey of IIM shows that about 60,000 crores rupees per year have been wasted in congestion [2]. Highway roads are being exploited in India. The expenditure in terms of road revenue generated is about India 35%, USA 96%, Japan 128%, and Germany 82%, respectively [3].

The paper is organized as follows: Sect. 2 stated the related work; in Sect. 3 the proposed model EEG is introduced. The Sect. 4 explored the issues and implementation of ICT in transport system. The paper is closed with conclusion and future scope.

2 Related Work

The research papers were explored and categorized related with congestion on roads, toll collection, and green roads. The papers on congestion on roads mainly dealt with the measures to control and detect congestion. Ranbaldi et al. (2012) analyzed the speed profile of Italian road network and proposed an algorithm for traffic control, provision, and governance. The algorithm was proposed for managing and monitoring the traffic, keeping in mind the safety of individual passenger [4]. Maria Pia Fanti proposed the combination of ICT (Information and Communication Technology) and ITS (Intelligent Transport System) which will prove attractive research in modeling, planning, managing, and control of traffic [5]. Dr. John Walker proposed a technology with the paradigm "MOVE AS YOU DRIVE" (PAYD) based on Time, Distance, Place (TDP) which was based on the calculation of road tax [6]. Manikonda P et al. proposed a technology named RFID that reduced the travel time of passengers. It also helped in tracing stolen cars, collecting toll taxes or road taxes. A database was created that is operated with the help of Dijkstra's algorithm [8]. Soomro proposed an algorithm for recognizing vehicle number in the form of an image called vehicle number recognition (VNR). This method was used to reduce the queues in toll tax by doing image processing on gray scale and then charging the toll according to the vehicle [9]. Danko J proposed a tool called General-Purpose Simulation System (GPSS) that evaluates the queuing in toll plaza. Also, it can analyze the decreasing and increasing staff to match with peak and off-peak demand [10].

3 Proposed Model—EEG

This paper basically deals with three new concepts. Among them the first one is E-roads that stands for electronic roads where the solar panels are implemented on roads to build up a smart highway, a smart parking lot, a smart playground, etc. In addition to E-roads, another new concept sums up in this paradigm to handle the road traffic intelligently called E-wallets (electronic wallet). The toll taxes should be levied on the vehicles moving on the roads depending on the time duration, and the utility of the road was used. The necessary amount will be deducted from electronic or smart card instead of waiting in the unnecessary queues when crossing the toll. Last but not the least, third model of this paper is G-roads (green roads) that inspires each and every individual to walk. We have proposed a model EEG in order to reduce traffic congestion on the roads and make our environment green and friendly as shown in Fig. 1.

Fig. 1 Proposed model EEG

3.1 E-Roads

E-roads highlight the concept of solar roadways which is already implemented in Idaho, Sandpoint. As the name suggests, the roads are solar powered panels enabled in place of concrete roads to make up a smart highway, a smart parking lot, a smart playground, a smart footpath, etc. [2]. The implementation of solar panels over roads could help to overcome the problem of greenhouse effect up to 75% [11].

3.1.1 Structure

The solar panels on roads are hexagonal in shape and each panel is linked up with other like molecules. It comprises the following layers as listed in the table below (Figs. 2 and 3).

Layer number	Layer name	Composition
Layer 1	tampered glass layer	Tampered glass consisting of the set of LED's
Layer 2	circuit layer	Consist of microprocessor to build the road intelligent
Layer 3	recycled waste	Consist of the recycled waste around the country

Fig. 2 Structure of solar roadways

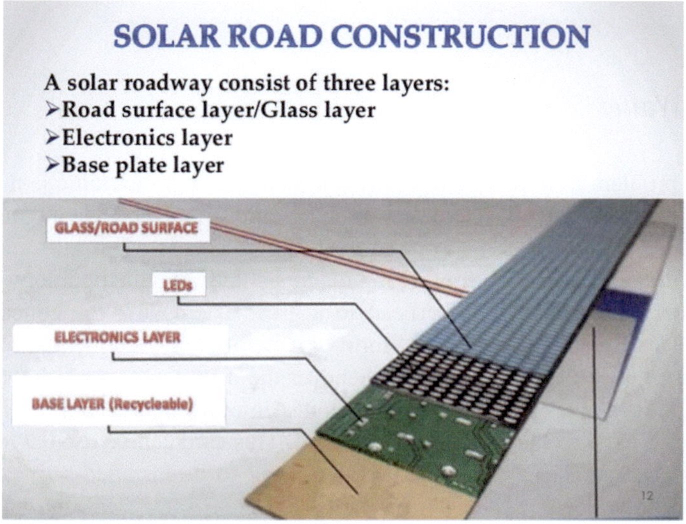

Fig. 3 Implementation of solar roadways [11]

3.1.2 Advantages of Solar Roadways

(a) *Intelligent highways*: As coal-fired and nuclear-enabled powered plants will diminish, the costs incurred in the generation of electricity can also be rolled back into the solar roadways. Each solar panel equipped with a microprocessor having its own address forms a wide area network (WAN). The installation of RFID (Radio Frequency Identification) tags on vehicles would help to track the location of vehicle.
(b) *Illuminated roads with intelligent LEDs*: The solar roadways designed should contain in-built LEDs which will "paint" the lanes and can be customized just in time of need. Illuminated highways would result in lesser accidents and safe driving in the night. The programming in the LEDs would be done in such a manner that the speed limit of the vehicles should be maintained, warning drivers instantly if driving too fast. The LEDs will also be used to paint the words right into the road, warning drivers of an animal on the road or a construction work or an accident.
(c) *Job Opportunities*: The construction of this system will create many job opportunities because a large number of skilled professionals will be required to design the EEG proposed model.
(d) *Energy Production*: Sun is the cleanest form of nuclear power. The energy reaching the Earth is more than 10,000 times of what is required. If even 1% of efficiency is achieved by us at low cost and if we store that energy, then we can use that energy which is enough for nine and a half billion people without causing pollution. The fossil fuels used in the generation of electricity can be reduced up to a great extent with the use of solar roadways.
(e) *Economic recovery*: By using the solar roadways, the nation can save three times the energy it is actually using presently.

3.2 E-Wallet

The main problem in Indian transportation system is about the traffic jam that ruins the life of the traveler be it small cities or metro cities. The vehicles moving through the toll booth have to spend an average of 10 min to cross it. This kills a lot of time of the traveler as well as result in a long queue behind. The methodology proposed in the paper can reduce this overhead to a great extent with the concept of the E-wallet. This E-wallet can be in the form of a credit card which would be credited to the user at user's expenses. Then whenever the user passes a toll booth or uses the roads, the amount can be deducted from it. A mobile app for mobile phones can be developed for e-recharge of the plastic card. This card can be used at toll booths and roads. The camera and special software need to be installed on heavy traffic roads. The utility of the card can be increased if the same card can be used for traveling by train or metro. The method is summarized in Fig. 4.

Fig. 4 Method of E-wallet

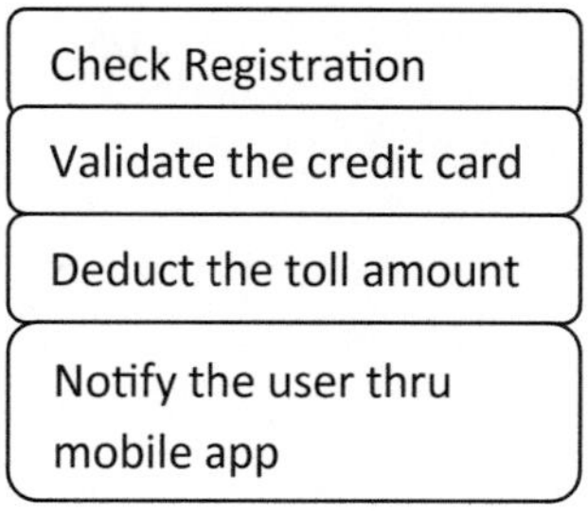

3.3 G-Roads

G-roads known as green roads can be termed as designer green roads attracting people to walk. As if a person wants to walk, then there are several environmental and health-related problems one has to face. The very first problem is that there is no separate path provided in the roads for pedestrians and cyclist (a person who prefer to travel through bicycle). Second problem is the width of roads that is reducing day by day, so it is difficult to provide extra lane for pedestrians. As discussed in E-wallet approach, the revenue collected by the people is somewhat extra, so government needs to use that amount in widening the roads and planting trees along the roads. The path provided for the pedestrians should have a covering to give protection from Sun. This simple approach will motivate huge amount of employment among the illiterate and unemployed crowd. The work involves widening of the roads and as a gardener in planting more and more trees. Facilitating roads with sheds along the roads can also be used to implant solar panels that are used in switching on the lights at night. This will also reduce the amount wasted in street lights. Health is the major issue faced by everyone in these days. Obesity, high blood pressure, heart problem, and diabetes are some of the current topics that add on in this list. Walking is the only solution that proves as a blessing of every disease.

4 Issues and Challenges in Implementing ICT-Enabled EEG Model

The issues and challenges in implementing EEG model are the costs as follows: the life style cost, infrastructure, and alternate fuel. We can associate major issues and challenges to the same as listed below:

(a) Cost: The implementation of the new technique would need a huge cost to be invested that could prove the main barrier from the government's end.
(b) Infrastructure: The conversion of infrastructure to a new system would raise many political and social eyebrows.

(c) Civic sense and traffic sense among people: People in India greatly lack the two qualities mentioned above. There is a need to train or educate people for the new change and adaptation to the new system.
(d) Restructuring of traffic rules: The new system would bring new traffic rules and the compulsion of the people to abide by the rules itself would be challenging. Until and unless the rules are strict and people obediently follow it, the new system would not flourish.
(e) Lack of standardization: National procedures and regulations concerning transport are not standardized for different states in India. Like, in Chandigarh, a black film on the car window is banned, whereas in NCR no such rule is followed.

5 Conclusion and Future Scope

The paper proposed a new ICT-enabled transport EEG model. The paper explored the existing problems of the transport system. The EEG model introduced the concept of E-roads, E-wallets, and G-roads. The implementation of this model would reduce the risk of road accidents, improvement in the present environment condition and traffic congestion economically.

This work can be further extended to the next level by the actualization of the implementation of the proposed EEG model and the new adaptions required can be addressed further.

References

1. Yu, Y.T., Lau, M.F.: A comparison of MC/DC, MUMCUT and several other coverage criteria for logical improving transportation requires a new solution: peter. J. Muller, p.e (2006)
2. Srikanth, R.P.: Can intelligent transport system solve india's traffic congestion problem? (2014)
3. Dado, M., Spalek, J., Janota, A.: Present and future challenges of ICT for intelligent transportation technologies and services. IEEE explore, pp. 107–110 (2009)
4. Rambaldi, S., Marchioni, M., Bazzani, A., Giorgini, B.: Global analysis on the whole italian road network. IEEE Conference Publication, pp. 1678–1682 (2012)
5. Black, W.R.: Socio-economic barriers to sustainable transport. J. Transp. Geogr. **8**, 141–147 (2000)
6. Banister, D., Stead, D.: Impact of information and communication technology on transport. Transp. Rev. **24**(5), 611–632 (2004)
7. Black, W.R.: Transportation: A Geographical Analysis. The Guilford Press, New York (2003)
8. Boyle, L.N., Mannering, F.: Impact of traveler advisory systems on driving speed: some new evidence. Transp Res Part C **1**, 57–72 (2004)
9. Manikonda, P., Yerrapragada, A.K., Annasamudram, S.S.: Intelligent traffic management system. IEEE Conference Publication-2011, pp. 119–122 (2011)
10. http://ieeexplore.ieee.org/xpl/mostRecentIssue.jsp?punumber=5161964
11. https://www.indiegogo.com/projects/solar-roadways#/story

Multi-criteria Rating Using Fuzzy Ranking for Improving Soil Recommendation System

Babita Chaudhary and Sandeep Dahiya

Abstract The fuzzy methodology in light of Multi-criteria Decision-Making (MCDM) system for selecting the ideal answer for enhancing creation in the field of agribusiness by selecting the ideal soil for a specific crop with a set of linguistic variables of the well-known AHP methods by the ratings of each dimension of fuzzy matrix and the overall rating of soil. The majority of the crops need very much depleted; salt-free soils and lean toward very much drained medium-textured soil with ideal physical properties; and impartial pH as soil has ideal physical, chemical, and natural properties; therefore soil suitability standard issues appear to be a nonappearance of capacity and assessment from every angle. Personalization advancements and recommenders framework help to beat the issues by giving customized proposals for farmers according to the suitability of data. The present investigation is utilized to calculate normalized weight using defuzzified value or crisp value and to create a rating of crops for the recommendation system.

Keywords Fuzzy linguistic variables · Fuzzy multi-criteria decision-making · Recommender systems · Fuzzy number (FN)

1 Introduction

An ideal support methodology is a key backing for generation in the agribusiness. Under good management practices in the agribusiness, the soils can be beneficially utilized for farming, plant, ranch crops, natural alteration, and green manure which build the soil richness status. The majority of the crops need very much depleted, salt-free soil, and lean toward very much drained medium-textured soil with ideal

B. Chaudhary (✉) · S. Dahiya
Department of Electronics and Communication Engineering, Bhagat Phool Singh Mahila Vishwavidyalaya, Khanpur Kalan, Sonipat 131305, India
e-mail: babita2ch@gmail.com

S. Dahiya
e-mail: sandy_dahiya2001@yahoo.com

A.K. Saini et al. (eds.), *ICT Based Innovations*, Advances in Intelligent Systems and Computing 653, https://doi.org/10.1007/978-981-10-6602-3_5

physical properties and impartial pH as soil has ideal physical, chemical, and natural properties. The soil suitability standard issues appear to be a nonappearance of capacity and assessment from every angle [1]. Suitable drainage facilities are obligatory in water-logged ranges which additionally reduce the further issues like saltiness and alkalinity. Fruitful crop management depends on selecting suitable products to the kind of soil present in that locale. So as to use sound judgment for the soil for any crop, it is commonly important to have a certain adequate measure of data.

Personalization advancements and recommender frameworks help to beat the issues on decision-making by giving customized proposals with respect to which data is most applicable to clients (farmers, customers) [2]. All of these aspects make sure that it is vital to use a model which more precisely give implicit or numerical scale value. In the present investigation, fuzzy multi-criteria decision-making is used to provide a model to improve the soil recommendation system. Further, fuzzy multi-criteria decision-making has prevalently used four different methods: (i) the out ranking, (ii) the utility theory and value based, (iii) the multi-object programming, and (iv) group decision and negotiation theory-based method [3].

The present investigation is organized as follows: Sect. 2 discusses the soil analysis for the suitability of soil and Sect. 3 presents linguistic variables for the soil multi-criteria decision-making and methodology of the devised system. Section 4 demonstrates the results and discussion following the concluding remarks with a future scope. The next section discusses the soil analysis for the suitability of soil.

2 Soil Analysis

The changed soil's suitability standard issue appears to be a nonappearance of capacity and assessment from every angle such as different territory variables; officers are inclined to have differing clarifications for such declarations as "great moisture holding limit", "low to direct fruitfulness", and "poor to direct yields". As a beginning move toward achieving an unrivaled understanding and more noticeable consistency of assessments, it gives off an impression of being vital, as it needs to

- Define the region, consider the center soil suitability;
- Specify for each region, compute different classes; and
- Rate each class according to its limitation for zone usage sorts and individual items.

The portrayal can be used to rotate soil as demonstrated by their suitability in like manner for particular unrests [4]. Soil contains chemicals in humorous amount and some of the soil criteria are depicted in Table 1. pH factor of soil is a measure of the acidity and alkalinity in soils. pH levels range from 0 to 14, with 7 being unbiased, underneath 7 acidic, or more 7 alkaline [5], and the pH factor for soil property is depicted in Table 2.

Table 1 Soil criteria

Soil chemicals	Soil physical aspects
(a) pH	(a) Texture
(b) Organic carbon	(b) Drainage
(c) Fertility	(c) Depth

Table 2 pH factor for the soil property

pH scale	Soil property
3.0–6.0	Strong acid
6.1–6.5	Weak acid
6.6–7.3	Neutral
7.4–8.0	Weak alkaline
8.1–9.0	Strong alkaline

Soil pH is viewed as an expert variable in the soil as it controls numerous compound procedures. It particularly influences plant supplement accessibility by controlling the synthetic types of the supplement such as a few variables impact soil pH as parent material, precipitation, local vegetation, harvest developed, nitrogen, and other corrosive shaping composts and natural material, blazing fossil energizes, watering system, etc. The decompositions of plant, dead parts of animals, root, living, and dead microorganisms are the main causes of the presence of Organic Carbon (OC) in soil, and soil microorganisms for energy mainly depend on OC. Organic matter in soil varies with turnover time, rate of decomposition, the presence of carbon content, and particle size as a heterogeneous, dynamic substance. The fertility is related to the capacity of soil to support a particular natural community of plants. The focus is mainly on physical and chemical aspects of soil in terms of the maximum practical level of productivity. Biological activity in soil can be overridden by addition of fertilizers to increase plant growth.

The relative quantity of distinct sized soil particles or the thin mineral particles in the soil has defined the texture [6]. Class contains the relative amount of sand, silt, and clay that is vital on which texture of soil depends and soil texture classes take their names from the particle size categories. The rate and amount of water movement across the surface as well as downward through the soil is known as soil drainage. Slope is a very vital factor in soil drainage and soil color is an indicator of soil drainage [7]. Well-drained soil has clear, bright color while poor drainage indicates mixed, drab, and dominantly gray color and low-lying are within the land scape receiving run-off water. The vertical distance from the surface to the soil to a layer which is vitally needed to stop the plant roots growth in the downward direction is called effective depth. Terms used to express soil depth from a layer that slows down the root development are (i) Very Shallow—less than 10 inches soil surface, (ii) Shallow—10–20 inches soil surface, (iii) Moderately deep—20–36 inches soil surface, (iv) Deep—36–60 inches soil surface, and (v) Very deep—60 inches or more soil surface. Deep soil has more amounts of water and plant nutrients in comparison with shallow soils with similar textures.

The succeeding section will provide a framework of linguistic variable for the soil multi-criteria decision-making and methodology of the devised system for present investigation.

3 Linguistic Variables for the Soil Multi-criteria

The over simplification of the decision problem could lead to few criteria being used. Often proxies must be used since the criteria of interest may not be determinable. The calculative criteria, objectives, and attributes should be identified with respect to the problem situation. The analyzing system is responsible for the set of attributes or criteria in the recommendation process. There are following calculative criteria to address the soil suitability decision-making as in chemical (fertility, pH, OC) and in physical (texture, drainage, depth) [8]. In this paper, criteria like pH factor, OC, and fertility are used as multi-attributes for improving the soil recommendation systems.

3.1 Framework of Devised Recommendation System

Fuzzy logic is useful for decision-making methods because of the ability to use fuzzy terms such as uncertainty, semi-truth, unclearness, etc. in fields of business, agriculture, and public services. There are a variety of inputs that are essential to take into concern on the output as discussed in the study. The evaluation of multiple rule boxes in MATLAB is required for the proposed problem of recommendation system process. The same can be done with the help of program via M-file. The fuzzy multi-criteria methods are shown in Table 3.

The devised system for present investigation is shown in Fig. 1, in which the approaches extend with certain amendment aiming to achieve more precise result to improve the performance of multi-criteria recommendation system. First, team assigns linguistic terms for each criterion which is converted into fuzzy numbers and creates the fuzzy decision matrix of assigning terms. Second, calculate the average

Table 3 Fuzzy multi-criteria methods

Features	Number of judgments	Ease to use	Trustworthiness
Rating	N	Low	Approximation
Ranking	N	High	Not precise
Pairwise comparison	n(n−1)/2	High	Quite precise
Trade-off analysis	<n	Medium	Quite precise

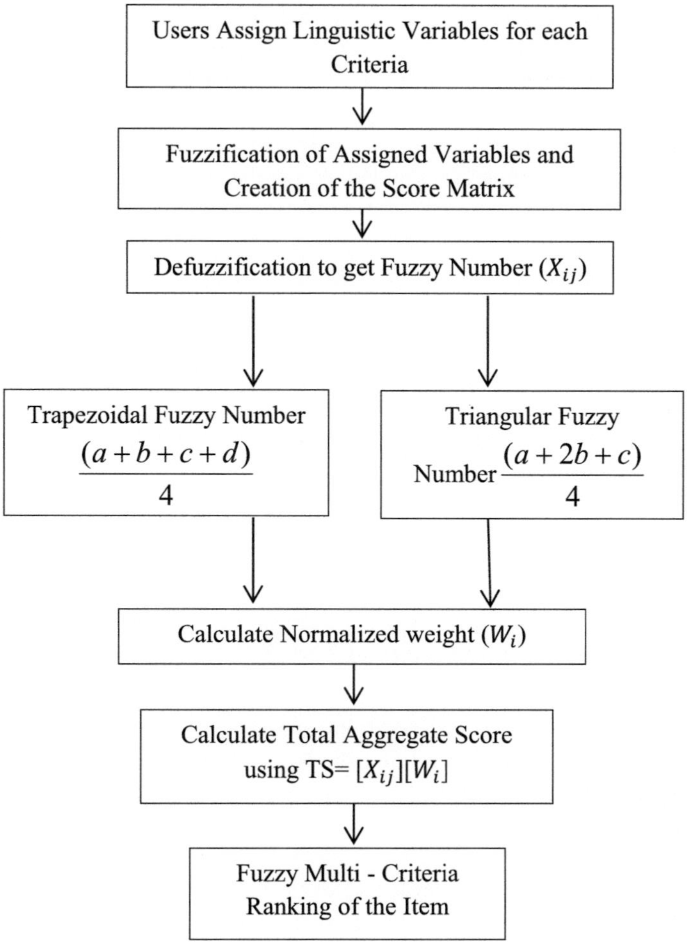

Fig. 1 Flow/Ebb diagram of the devised system for the present investigation

scores of fuzzy with help of the same, defuzzified values, and normalized weight for each criterion is evaluated, then total aggregated score for crop against each criterion is calculated and finally evaluate the overall ratings for the final outcome.

3.2 Fuzzy Linguistic Modeling

Fuzzy logic provides descriptive and qualitative form for vague concepts. A set in which the degree of membership in a set is given by a number is between 0 and 1, known as fuzzy set. A function, called membership function, defines the fuzzy set. The elements in input space domain are mapped to their membership value in the set by membership function. Let a fuzzy set be B, then the membership function of B is defined as μ_B and the membership value of x is μ_B (x). A linguistic variable carries a qualitative value, i.e., using linguistic term or a quantitative value for using a corresponding membership function [9]. A membership function for a fuzzy set B in X is defined as $\mu_B(x) : x \in X \rightarrow [0, 1]$, in which X is domain space. According to the context in which x is used, the fuzzy membership function $\mu_B(\mathrm{x})$ has different interpretations. For example, the membership value of a soil x in the fuzzy set of team preference can be calculated by the team's degree of preference in that soil. The membership function defines the intensity of team's preference in favor of crop x. Fuzzy set theory allows a continuous value for $\mu_B(\mathrm{x})$ between 0 and 1 as given below:

$$\mu_B(x) = \begin{cases} 1 \quad \mathit{iff}\ xB \\ 0 \quad \mathit{iff}\ x\ not\ B \\ 0<p<1, \mathit{if}\ x\ partially\ belongs\ to\ B \end{cases}$$

3.3 Fuzzy Multi-criteria Team-Crop Rating Matrix

In the present work, instead of specifying numerical scale, the user's relevance feedback is collected by use of the linguistic terms. The ratings of team are fuzzified using triangular membership functions, supplied to decide the degree of membership for the fuzzy set of user's preferences [10, 11]. Let the fuzzy variable degree of recommendation of soil based on a criteria consists of fuzzy values and are represented by using five linguistic terms: {Not Preferable (NP), Less Preferable (LP), Fairly Preferable (FP), Preferable (PR), and Highly Preferable (HP)} [12, 13]. The terms of the fuzzy variable degree of recommendation have membership function. Triangular fuzzy number is assigned to the linguistic variables as shown in Table 4.

The next section of the investigation demonstrated the results and discussion of the study under reference.

Table 4 Linguistic variables with TFN values

Linguistic variables	TFN
Not preferable (NP)	(0, 0, 0.20)
Less preferable (LP)	(0 ,0.20, 0.40)
Fairly preferable (FR)	(0.20, 0.40, 0.60)
Preferable (PR)	(0.40, 0.60, 0.80)
Highly preferable (HP)	(0.60, 0.80, 1)

4 Result and Discussion

The element of rating in the matrix is given as $R_i t_j = \mu_A^{C_t}(I_j)$, in which $\mu_A^{C_t}(I_j)$ is the membership value on preference fuzzy set A of crop I_j under the criteria C_t of team i. In this investigation, C_k represents the multiple criteria and contains fertility, OC and pH_factor, where k = 3. Each matrix element has the fuzzy multi-criteria rating representation of team i for crop j which is represented as $(\mu^{c1}(Ij), \mu^{c2}(Ij), \mu^{c3}(Ij))$. A typical fuzzy multi-criteria team-crop rating matrix representation is given in Table 5.

To measure the accurate results, the normalized weight is evaluated using defuzzification of fuzzy numbers as simple team ratings which are not exact or not accurate so the linguistic approach of fuzzy is used to rate the preferences to the team. The main objective is the selection of the highly suitable for accurate crop from m different crops on the behalf of k multi-criteria $(c_1, c_2, c_{3,\ldots,} c_k)$. The rating is considered as R_{itj}, which is assigned to alternative crops I_j by the U_i teams on the C_t criteria [9, 14–17]. Then, the average of fuzzy numbers of the different ratings will be

$$R_{itj} = \frac{1}{p} \otimes (R_{it1} \oplus R_{it2} \oplus \cdots \oplus R_{itp}), \tag{1}$$

p = 1, 2, 3…p.

The average fuzzy scores matrix is obtained by using the equation for triangular fuzzy numbers as shown in Table 6:

Table 5 Fuzzy multi-criteria (fertility, OC, and pH) team-crop ratings matrix

	Crop I_1 $C_1C_2C_3$	Crop I_2 $C_1C_2C_3$	Crop I_3 $C_1C_2C_3$	Crop I_4 $C_1C_2C_3$
Team U_1	(0.60, 0.80, 1) (0, 0, 0.20) (0.40, 0.60, 0.80)	(0, 0.20, 0.40) (0.60, 0.80, 1) (0, 0.20, 0.40)	(0.40, 0.60, 0.80) (0.40, 0.60, 0.80) (0.40, 0.60, 0.80)	(0.40, 0.60, 0.80) (0, 0.20, 0.40) (0.20, 0.40, 0.60)
Team U_2	(0.60, 0.80, 1) (0,0.20, 0.40) (0.40, 0.60, 0.80)	(0.20, 0.40, 0.60) (0.40, 0.60, 0.80) (0, 0.20, 0.40)	(0.60, 0.80, 1) (0.20, 0.40, 0.60) (0.60, 0.80, 1)	(0, 0, 0.20) (0.20, 0.40, 0.60) (0, 0.20, 0.40)
Team U_3	(0.40, 0.60, 0.80) (0, 0.20, 0.40) (0.60, 0.80, 1)	(0, 0, 0.20) (0.40, 0.60, 0.80) (0, 0, 0.20)	(0.20, 0.40, 0.60) (0.20, 0.40, 0.60) (0.20, 0.40, 0.60)	(0,0, 0.20) (0.60, 0.80, 1) (0.20, 0.40, 0.60)
Team U_4	(0.20, 0.40, 0.60) (0, 0, 0.20) (0.60, 0.80, 1)	(0.20, 0.40, 0.60) (0.40, 0.60, 0.80) (0, 0.20, 0.40)	(0.60, 0.80, 1) (0, 0.20, 0.40) (0.40, 0.60, 0.80)	(0, 0.20, 0.40) (0, 0.20, 0.40) (0.20, 0.40, 0.60)
Team U_5	(0.60, 0.80, 1) (0.20, 0.40, 0.60) (0.40, 0.60, 0.80)	(0, 0.20, 0.40) (0.60, 0.80, 1) (0, 0, 0.20)	(0.40, 0.60, 0.80) (0.20, 0.40, 0.60) (0.40, 0.60, 0.80)	(0.20, 0.40, 0.60) (0.20, 0.40, 0.60) (0.20, 0.40, 0.60)

Table 6 Average fuzzy score matrix

	Crop I_1	Crop I_2	Crop I_3	Crop I_4
C_1	0.48, 0.68, 0.88	0.08, 0.24, 0.44	0.44, 0.64, 0.84	0.12, 0.24, 0.44
C_2	0.04, 0.16, 0.36	0.48, 0.68, 0.88	0.2, 0.4, 0.6	0.2, 0.4, 0.6
C_3	0.48, 0.68, 0.88	0, 0.12, 0.32	0.4, 0.6, 0.8	0.16, 0.36, 0.56

$$e = \frac{(a + 2b + c)}{4}. \tag{2}$$

Normalized weight for each criterion is calculated by division of the defuzzified score of individual criterion by the total criteria which are presented in Table 7 and also stored in a matrix W_j to calculate the final score for the crops as shown in Fig. 2.

Total aggregated for the final score (FS) for crops against each criterion is evaluated by using a simple additive method between the defuzzified X_{ij} and normalized weight for each criterion W_j as shown below:

$$\text{FS} = [X_{ij}][W_j]. \tag{3}$$

Total score for crops $(I_1)\,(I_2)\,(I_3)\,(I_4)$ is achieved from Eq. 3 and the final ranking of the crops is as provided in Table 8.

Table 8 demonstrates the final score for each crop with the help of normalized weight W_j by comparing the average values of crop with respect to the criteria as presented in Table 6. Figure 3 shows the score of crops for the soil suitability which will also decide the rank of degree of interest of recommendation.

The succeeding section will explore the concluding remarks of the present investigation following the future scope of the work.

Table 7 Normalized weight for the multi-criteria using defuzzified values

Crops	Criteria	Avg. Fuzzy values	Defuzzified values	Normalized weights
Crop I_1	C_1	0.48, 0.68, 0.88	0.68	0.3333
Crop I_1	C_2	0.04, 0.16, 0.36	0.18	0.3214
Crop I_1	C_3	0.48, 0.68, 0.88	0.68	0.3333
Crop I_2	C_1	0.08, 0.24, 0.44	0.25	0.3289
Crop I_2	C_2	0.48, 0.68, 0.88	0.68	0.3333
Crop I_2	C_3	0, 0.12, 0.32	0.14	0.3181
Crop I_3	C_1	0.44, 0.64, 0.84	0.64	0.3333
Crop I_3	C_2	0.2, 0.4, 0.6	0.40	0.3333
Crop I_3	C_3	0.4, 0.6, 0.8	0.60	0.3333
Crop I_4	C_1	0.12, 0.24, 0.44	0.26	0.3250
Crop I_4	C_2	0.2, 0.4, 0.6	0.40	0.3333
Crop I_4	C_3	0.16, 0.36, 0.56	0.36	0.3333

$$\begin{array}{c} \\ C_1 \\ C_2 \\ C_3 \end{array} \begin{array}{cccc} I_1 & I_2 & I_3 & I_4 \\ \end{array}$$

$$\begin{matrix} C_1 \\ C_2 \\ C_3 \end{matrix} \begin{bmatrix} 0.68 & 0.25 & 0.64 & 0.26 \\ 0.18 & 0.68 & 0.40 & 0.40 \\ 0.68 & 0.14 & 0.60 & 0.36 \end{bmatrix} \begin{bmatrix} 0.3333 \\ 0.3214 \\ 0.3333 \end{bmatrix} W_j$$

Fig. 2 Matrix to calculate final score of crop

Table 8 Final ranking of the crops

Crops	I_1	I_2	I_3	I_4
Final Scores	0.5112	0.3486	0.5419	0.3352
Rank	2	3	1	4

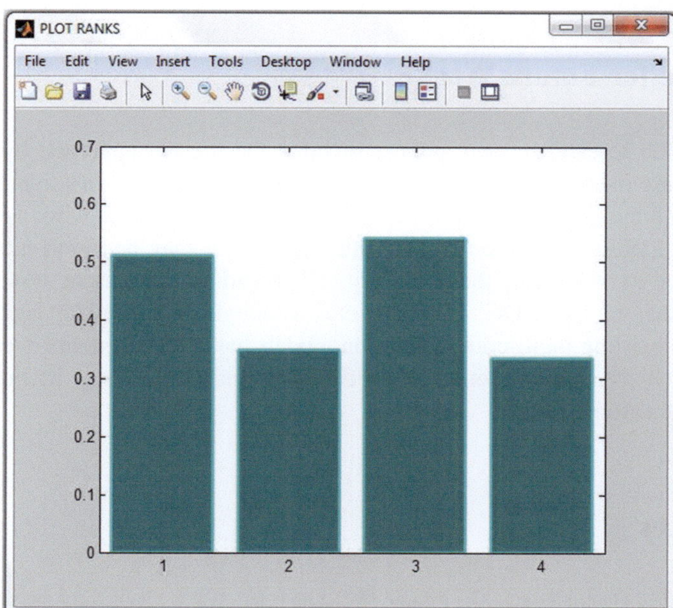

Fig. 3 Rank plot of crops for the present investigation

5 Conclusion

As the agriculture is moving toward usage of technology in extremely diverse and rich environments, the related cooperation of agri-field about the farmer in the decision-making and recommendation process has a major interest, so building an intelligent system that can better predict and anticipate the needs of farmers. In the present investigation, a new system has been presented that allows the intuitive inclusion of estimation imprecision in the analysis and ranking of soil for a recommendation. An attempt is made toward the finding of an optimal relation between the overall ratings and individual criteria ratings in multidimensional datasets. The crops (I_1) (I_2) (I_3) (I_4) have taken for the present investigation and the results for the same provided the score of individual crops for soil suitability which has also provided the rank of degree of interest of recommendation as the rank 2 for I_1, rank 3 for I_2, rank 1 for I_3, and rank 4 for I_4, respectively as depicted in Table 8 and Fig. 3. The experimental study has been conducted on the dataset of soils and experimental results confirmed that ratings in multi-criteria have been efficiently leveraged to improve the accuracy of recommendation on a real-world dataset.

6 Scope for Future Work

In future, the researchers and authors broaden the present approach by utilizing fluffy mixture methodologies and other machine inclining strategies to enhance the precision of the framework. The methodology in the present work is centered around sorts of soil suitable for a specific yield but it may be worth meaning for future scope to investigate the practicality of extending the present system to different circles, e.g., pH, OC, Minerals, etc. Although incorporating fuzzy set the oryinto the existing multi-criteria decision models has certain benefits over existing models such as better account for uncertainty and subjectivity in outsourcing vendor evaluation etc.

References

1. Addeo, G., Guastadisegni, G., Pisante, M.: Land and Water Quality for Sustainable and Precision. I World Congress on Conservation Agriculture, Madrid (2001)
2. Sałabun, Wojciech.: Application of the Fuzzy Multi-criteria Decision-Making Method to Identify Nonlinear Decision Models. Int. J. Comput. Appl. **89**(15), 1–6 (2014)
3. Ma, Jun., Zhang, Guangquan., Jie, Lu.: A fuzzy hierarchical multiple criteria group decision support system-decider and its applications. Springer Fuzziness and Soft Computing **267**, 383–403 (2011)
4. Santos, F.J.J., Camargo, H.A.: Fuzzy Systems for Multicriteria Decision Making. CLEI Electronic J. **13**(3), 1–8 (2010)

5. Kihoro, J., Bosco, N.J., Murage, H.: Suitability analysis for rice growing sites using a multicriteria evaluation and GIS approach in great mwea region, Kenya. Springer **12**(3), 1–24 (2013)
6. Beek, K.J., Burrough, P.A., McCormack, D.E.: Quantified land evaluation procedures. J. Plant Nutr. Soil Sci. **151**(1), 74 (1988)
7. Burrough, P.A.: Fuzzy mathematical methods for soil survey and land evaluation. J. Soil Sci. **40**(3), 477–492 (2006)
8. Komatsuzaki, Masakazu., Ohta, Hiroyuki.: Soil management practices for sustainable agro-ecosystem. Springer Sustainability Sci. **2**(1), 103–120 (2007)
9. Bellman, R.E., Zadeh, L.A.: Decision-making in a fuzzy environment. Manage. Sci. **17**(4), 141–164 (1970)
10. Deng, Hepu.: Multi-criteria analysis with fuzzy pairwise comparisons. IEEE Trans Fuzzy Syst. **2**, 726–731 (1999)
11. Kabir, G., Hasin, M.A.A.: Comparative analysis of AHP and fuzzy AHP models for multicriteria inventory classification. Int J. Fuzzy Logic Syst **1**(1), 1–16 (2011)
12. Wang, Z.X., Mo, Y.N.: Ranking Fuzzy numbers based on ideal solution. Fuzzy Inf Eng. **2**(1), 27–36 (2010)
13. Edwards, J.H., Wood, C.W., Thurlow, D.L., Ruf, M.E.: Tillage and crop rotation effects on fertility status of a Hapludalf soil. Soil Sci. **56**, 1577–1582 (1999)
14. Liu, Xin-Wang., Han, Shi-Lian.: Ranking Fuzzy numbers with preference weighting function expectations. Elsevier Comput. Mathematics Appl. **49**, 1731–1753 (2005)
15. Buckley, J.J.: Fuzzy hierarchical analysis. Uncertainty in Risk Assessment, Risk Management, and Decision Making Advances in Risk Analysis **4**, 389–401 (1987)
16. Palanivel, K., Sivakumar, R.: Fuzzy multicriteria decision-making approach for collaborative recommender systems. Int J Compu Theory Eng. **2**(1), 57–63 (2010)
17. Boender, C.G.E., de Graan, J.G., Lootsma, F.A.: Multi-criteria decision analysis with fuzzy pairwise comparisons. Fuzzy Sets Syst. **29**, 133–143 (1989)

Design and Development of the Agricultural Model: A Way to Connect Farmer Community to Agriculture Market for Betterment of Rural Development

Tejas Ghadiyali, Kalpesh Lad and Jayesh Dhodiya

Abstract Rural development is a process of sustainable improvement in the quality of life of rural people especially farmer community and overall development of rural areas. Agricultural development constitutes the crucial aspect of rural development. Agricultural development is possible through the betterment of farm activities and agri-business activities. In this study, the researchers have designed and developed the agricultural model that predicts future market price of agriculture commodities at the time of crop cultivation and thereby connects the farmer community to the agriculture market. The farmer community will be able to get the knowledge about future market price of agriculture commodity and accordingly plan for crop cultivation, crop production and other farm management activities. Thus, such generated knowledge from this model helps the farmer community in its decision making at the time of crop cultivation and reducing risk in their agri-business and thereby helps in rural development.

Keywords Agricultural model · Rural development · Agri-product price prediction

T. Ghadiyali (✉)
UCCC&SPBCBA&UACCAIT, Veer Narmad South Gujarat University, Surat, Gujarat, India
e-mail: tejas_ghadiyali@rediffmail.com

K. Lad
SRIMCA, Uka Tarsadiya University, Bardoli, Gujarat, India
e-mail: kalpesh.lad@utu.ac.in

J. Dhodiya
Department of AM&H, S.V. National Institute of Technology, Surat, Gujarat, India
e-mail: jdhodiya2002@yahoo.com

A.K. Saini et al. (eds.), *ICT Based Innovations*, Advances in Intelligent Systems and Computing 653, https://doi.org/10.1007/978-981-10-6602-3_6

1 Introduction

The rural development generally refers to the process of improving the quality of life and economic welfare of people living in villages. Rural development is a comprehensive and multidimensional concept, and encompasses the development of agriculture and allied activities and, above all, human resources in rural areas. "*Agricultural research has played a vital role in agricultural transformation and in reducing hunger and poverty and its role in the Twelfth Plan will be crucial*" [1].

India is predominantly an agrarian economy with 70% of its total area falling under rural category [1]. As a crucial aspect of rural development, the agriculture sector in India plays a significant role in the economic development of India. More than 60% workforce of the total population of India is engaged in this sector and 15.2% GDP of overall economy of India is accounted from this sector [1]. These facts show the importance of the Agriculture sector in the rural development of India. *The total budgetary allocation for all rural development programmes by the Government of India in the Eleventh Plan was 2,91,682 crores which accounted for 25% of the total Central Budget Plan provision* [2]. This fact shows the importance of Rural Development for Government of India.

In this context, the researchers have been developing an 'Agriculture Intelligence' [3] system to bridge the gap between information provided to the agriculture stakeholder and the real knowledge requirement of this community, and this study is a part of it. This study represents the design and development of the agricultural model, which suggests a way to bridge the gap between the farmer community and the agriculture market, and thereby enhance their ultimate goal of profit making in agri-business. Mathematical model is a representation of a physical phenomenon, system or device of a real life into mathematical form. It also considerably contributes for various farm management activities such as 'agricultural crop production techniques', 'crop production inputs', 'crop growth', 'crop production quality improvement' and 'controlling risk for an agricultural harvest'. Some of the mathematical models of activities such as 'Compost Pile Temperature Prediction' [4] and 'Pest Management' [5] are also useful in betterment of farm management. Due to complexities of these mathematical models the researchers could concentrate only on particular aspects of farm operations at a time. Information and communication technology also play a vital role in the development of modern agriculture and contributes significantly in the market price prediction of agriculture commodity [6, 7].

This paper is segregated into six sections. The next section describes literature review of this study followed by Design of the Agricultural Model in the Third section. The fourth section elaborates development part of this model including Model Construction, Model Validation, and Knowledge Generation as sub section. The fifth section explains the proposed model's Analysis and Result in the form of tables and charts followed by conclusion in the sixth section.

2 Literature Review

Many researchers have been carried out in the domain of agriculture sector for the betterment of rural management. The research may relate with the farm activity, farm extension activities, agricultural planning, agricultural economics, and up-gradation of agri-business.

Haciyev [8] has proposed a model which helps the investors in agro industrial company of Moscow, Russia. According to the author of this study, the given method is expedient for using in aggregate with arrangement process priority and risk-component of projects. Ghamari et al. [9] have proposed a mathematical model Predicting the Terminal Velocity of Chickpea, Rice and Lentil in Tehran, Iran. The results showed that size and moisture content have significant effects on terminal velocity in each of the three crops. Allen and Schuster [10] developed a robust mathematical model to optimize the risk on an agriculture harvest. Bai et al. [11] presented a stochastic and deterministic model for agriculture production networks. Through this model they have demonstrated how diseases creep in and the resulting network vulnerabilities were analyzed.

The volatility of an agriculture commodity price is very high and therefore price forecasting for decision makers in this domain has become more sensitive and challengeable. Several studies were also conducted in the domain of agriculture price prediction. Rao et al. [12] studied the seasonal variation and forecasting in wholesale price of rice in Guntur district of Andhra Pradesh. Ticlavilca et al. [13] have performed multiple predictions for agricultural commodity prices before 1, 2 or 3 months. Hoffman [14] has developed a model to forecast the seasonal average of an agriculture commodity 'Corn' using future prices. In this study the researcher has used parameters such as basis value, marketing weights and composite of monthly futures and cash prices to forecast the seasonal average U.S. corn farm price. The result of this study was compared with the U.S. Department of Agriculture's projection on World Agricultural Supply and Demand Estimates and statistically no significant difference was found. The results of this model are also reliable and timely. Kaur et al. [15] have discussed the applications and techniques of Data mining in agriculture considering the problem of price prediction of crops. To increase the accuracy percentage of price prediction, the researchers utilized back propagation neural network prediction model using the Coimbatore market price of tomato as an example and simulated the result in MATLAB. Huang et al. [16] checked the role of high-frequency data in forecasting agricultural commodity futures. They used the Realized GARCH model developed by Hansen, Huang and Shek (2012) to estimate and forecast price volatility for agricultural commodity futures. The researchers also compared results and argued that Realized GARCH models perform better than GARCH and EGARCH models in both fitting and forecasting endeavors and also show the advantages of the Realized GARCH model.

Other studies also show the utility of ICT in various farm management activities [17, 18] and thus help to farmer community and there by contributes in rural development. The next section explains the design of the proposed model.

3 Design of the Agriculture Model

The design of the proposed agriculture model in the form of architecture is as given below. This architecture has five major components as (i) Data Source, (ii) Preprocessing and Data Storage, (iii) Knowledge Extraction, (iv) Knowledge Presentation, and (v) Stakeholder. Each component receives input from its earlier component, processes it and transfers resultant output to its next component for further process in the system (Fig. 1).

Data Source: This system utilizes various heterogeneous data sources such as agriculture product 'Market Price' and 'Arrival' data from APMC (Agricultural Producers Market Committee) portal [19] on commodity basis. The selected training data period is from 2009 to 2013 and the testing data period is calendar year 2014. The Data Source accepts Meteorological data from Indian Meteorological Department [20] and other weather portal [21] for the period 2004–2014.

Preprocessing and Data Storage: This component performs the cleansing task on the available data from the heterogeneous data sources. It performs the task in the sequence of (i) Data Cleaning, (ii) Data Integration, (iii) Data Transform and

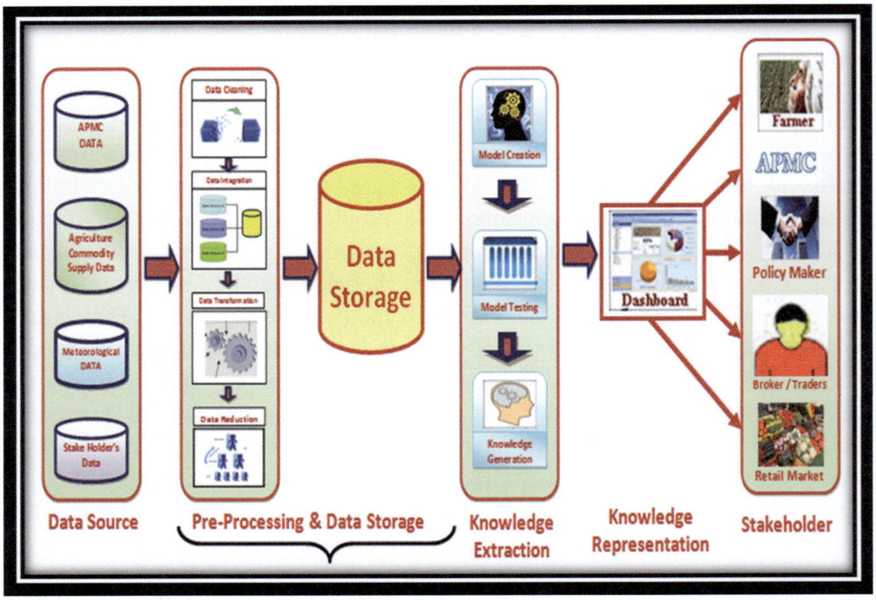

Fig. 1 The agricultural model architecture design

(iv) Data Reduction. After performing all such preprocessing tasks, the preprocessed data is transferred to data storage component for further analysis.

Knowledge Extraction: It is a crucial component of this system. It is classified into three sub components, such as (i) Model Construction, (ii) Model Validation and (iii) Knowledge Generation. Model Construction creates the price prediction model for selected agriculture commodity based on the selected training data period. Such generated model will be validated in the next subcomponent of Model Validation. It validates the model on the selected testing period data and offers accuracy to the model. The third subcomponent predicts future market price of agriculture commodity at the time of cultivation of that crop. This predicted price helps farmer community in their future risk management related to cultivated crop.

Knowledge Representation: This component is useful to present the generated knowledge from the previous component in the form of GUI display for the visual convenience of the stakeholders.

Stakeholder: This component contains various communities related to the agriculture sector such as Farmer, APMC, Agriculture Product Broker, Trader, Price Policy Maker or Agriculture Market Retailer. The next section elaborates the model development process.

4 Development of the Agricultural Model

Knowledge Extraction is the most crucial part of the overall design of this model. In Knowledge Extraction part we have a sub section as Model Construction, Model Validation, and Knowledge Generation. Each subsection has its own importance in the sequential process resulting into valuable ‘Knowledge’ in the form of predicted price of agriculture commodity prior to agri-business cycle process. Such facts are important for agriculture stakeholders in their decision-making.

4.1 Model Construction

A model is a conditional image of the researched object, designed to simplify the investigation. Agricultural model simulates on the basis of various meteorological and economical parameters. Daily Average Temperature, Humidity, and Rainfall act as meteorological parameters and Old Supply acts as an economical parameter to predict the agriculture commodity future market price. All the data that were used in this proposed model were gathered from the heterogeneous data source. The researchers have preprocessed ‘arrived data’ and generated dataset by matching date wise meteorological values in front of ‘agriculture arrival’ and ‘price’ data. Last decade Data from January 1, 2004 to December 31, 2013 was taken as training data and data from January 1, 2014 to December 31, 2014 was taken as testing data to validate the proposed model. The researchers applied three different approaches such as ‘Yearly’, ‘Seasonal’ and ‘Monthly’ to generate the model and validate it thereafter. Experimentally the researchers found that 'Monthly' approach has higher power to generate accurate results compared to other two approaches.

In this proposed model, the Market Price (Y) will be dependent variable and Daily Average Temperature ($x1$), Humidity ($x2$), and Rainfall ($x3$) as well as Old Supply ($x4$) will be independent parameters. Using such row information, the researchers generated a multiple linear regression model as per given below mathematical formula using *self-developed source code*. As shown in Eq. (1), 'Y' will be the predicted value and '*const*' is the intercept of the regression line. $\beta 1$, $\beta 2$, $\beta 3$ and $\beta 4$ are regression coefficients with respect to the parameters Temperature, Humidity, Rainfall and Old Supply, and their value is calculated as per the given equation in the following mathematical description of the proposed model.

$$Y = \sum_{i=1}^{n} \beta i * xi + \text{ Const} \tag{1}$$

$$Y_{(\text{predicted})} = Y + (\alpha - 1) * y + e \tag{2}$$

where

α	Fitness Factor
Y	Price
$x1$	Temperature
$x2$	Humidity
$x3$	Rainfall
$x4$	Old Supply
Const	$\bar{y} - (\beta 1 * x1 + \beta 2 * x2 + \beta 3 * x3 + \beta 4 * x4)$ (Intercept of the Regression Line)
βi	$\frac{\sum (xi*y) - \frac{(\sum xi) - (\sum y)}{n}}{\sum (xi^2) - \frac{(\sum xi)^2}{n}}$
e	Non-considerable parameters influence error
n	4

4.2 Model Validation

The performance of the model was evaluated by comparing the outputs of the model with the actual data. The researchers did not achieve enough amount of accuracy using only Eq. (1) (MLR) and hence experimentally introduced a 'Fitness coefficient-α' and generated the Eq. (2) for prediction of agriculture commodity market price. Here 'e' represents the error in prediction due to the factors not considered in this model. Fitness Coefficient-α has been calculated using the following steps.

Steps used to calculate Fitness Coefficient-α:

1. Initially started evaluating 'starting' and 'ending' boundaries of the experiments dataset. Considering the past agriculture commodity market price data, the researchers decided to start experiments from starting boundaries 0.5 and end with the ending boundary 3.0 with a step value of 0.05 and generated experiments data set values.
2. For every Individual fitness computation, used training samples for that respective month and tested the data accordingly and then, calculated individual learning error *E*. The formula is as follows:

$$E = \sum_{i=1}^{m} Ei, \quad \text{there into,} \quad Ei = |Yi - Y'i| \,/\, Yi$$

   Here, *m* is the number of validating sample; Ei is the ith sample error, i.e., the absolute difference between actual value Yi and predicted value Y′i divided by actual value Yi of the *i*th sample rounding at nearest integer value completely divisible by 25.
3. After having the individual fitness computation and learning error *E*, the mean absolute percentage error and accuracy percentage for an individual experiment dataset value was calculated as follows:

$$\text{MAPE} = E \,/\, m * 100 \quad \text{and} \quad \text{Accuracy}_{\text{per}} = 100 - \text{MAPE}$$

   Stored such accuracy percentage in the form of resultant value in resultant dataset.
4. Continued step 2 and step 3 for all the experiments dataset values and generated resultant dataset consisting of accuracy percentage values.
5. Then found out the maximum value (that is actually equilibrium trade-off value) from the resultant dataset, after and before this point in experiment data set, the accuracy value in resultant dataset is always lower. Accordance value of this resultant data set value in experiments data set value is known as the Fitness Coefficient-α.

The Fitness Coefficient-α determines price rise and thereby reduce the gap between predicted price and actual agriculture market price during that selected period. The next section is knowledge generation for the agriculture stake holders.

4.3 Knowledge Generation

In this third phase of the model development, the predefined Eq. (2) with Fitness Coefficient-α predicts future market price for agriculture commodity and given period by agriculture stake holders. For future date, the meteorological parameter values are calculated as the average of that day in the past decade. Old supply is considered for the last year data and these parameters value are applied in Eq. (2) to predict the future market price for the agriculture commodity in present. So this future market price is a 'generated knowledge' for the agriculture stakeholders and they may utilize the same for their decision making in their farm activities. The farmer community can also avail this 'generated knowledge' at the time of crop cultivation and take benefit of this future price to have precise inputs and other needed materials for crop production. The researchers also performed prediction testing for the month of July 2015 and the generated result in the form of accuracy for the different agriculture commodity is shown in the next section.

5 Result and Analysis

Table 1 shows the model experiments accuracy results of ten commodities using three different approaches yearly, seasonal and monthly and found that monthly approach gives better accuracy than that of other two approaches. User perform the analysis for ten agriculture commodities namely bhindi (ladies finger), brinjal, cabbage, cauliflower, green chili, lemon, onion, potato, surat beans (papdi) and tomato. But for visual convenient of reader, the researchers have shown only two commodities accuracy comparison chart as show in Figs. 2 and 3. This chart shows the comparative analysis of actual market price versus predicted market price for

Table 1 Accuracy analysis with usage of fitness factor–α

Model accuracy analysis with fitness factor						
Agriculture commodity	Yearly		Seasonal		Monthly	
	Accuracy (%)	Fitness factor	Accuracy (%)	Fitness factor	Accuracy (%)	Fitness factor
Bhindi (Ladies Finger)	74.21	1.15	76.81	1.25	83.67	1.35
Brinjal	56.05	2.85	63.33	0.92	82.42	1.26
Cabbage	56.85	1.00	64.74	0.97	86.36	1.15
Cauliflower	70.85	2.65	78.69	1.03	86.95	1.26
Green Chilli	74.75	2.65	80.07	1.63	82.95	1.66
Lemon	82.44	1.00	69.74	1.45	84.87	1.68
Onion	71.76	1.20	80.27	1.67	84.88	1.70
Potato	79.05	1.95	83.27	1.87	90.93	2.06
Surat Beans (Papdi)	74.00	0.85	77.11	1.23	84.21	1.41
Tomato	59.45	1.50	68.11	1.20	80.84	1.53

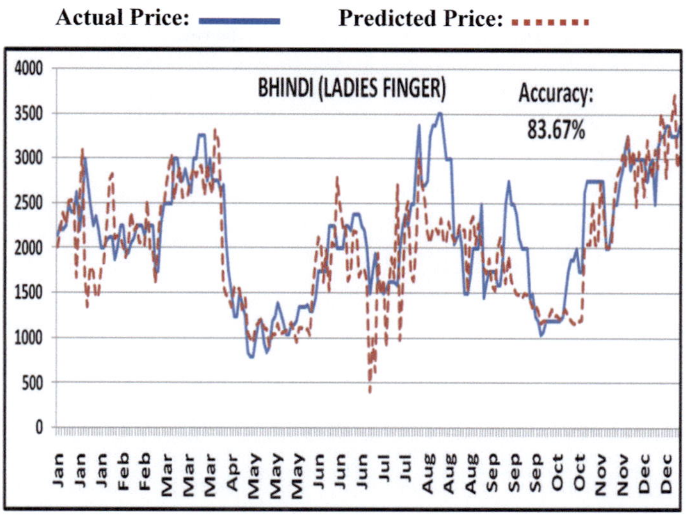

Fig. 2 Market price analysis: Ladies Finger

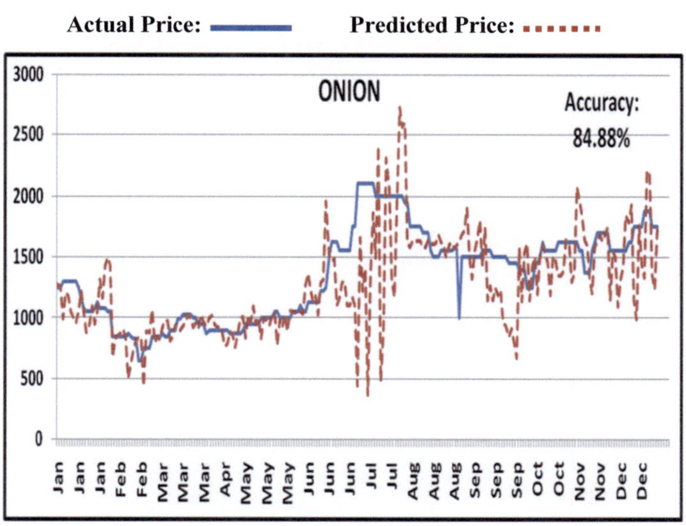

Fig. 3 Market price analysis: Onion

bhindi (ladies finger) and onion respectively. Tables 2 and 3 show the monthly average, maximum and minimum accuracy with fitness factor for that month for bhindi (ladies finger) and onion. Table 2 shows that the bhindi (ladies finger) model works better in the month of Dec, Feb and Mar and Table 3 shows that the onion models works better in the month of Aug, May, Apr, and Mar in the entire year successively.

Table 2 Monthly accuracy: Bhindi (Ladies Finger)

Bhindi (Ladies Finger)				
Model name	Avg accuracy	Max accuracy	Min accuracy	Fitness factor
BH_Jan	80.14	99.52	49.02	1.60
BH_Feb	90.96	99.72	79.38	1.55
BH_Mar	90.18	99.33	78.29	1.70
BH_Apr	81.06	97.12	59.04	1.00
BH_May	85.66	99.52	71.14	0.85
BH_Jun	81.79	98.93	60.80	1.65
BH_Jul	73.17	99.56	26.90	1.30
BH_Aug	77.66	99.16	53.54	1.50
BH_Sep	80.14	99.20	62.56	1.45
BH_Oct	83.09	99.43	57.85	0.95
BH_Nov	89.07	99.15	73.98	1.20
BH_Dec	91.13	99.39	75.73	1.50
Average	83.67	99.17	62.35	1.35

Table 3 Monthly accuracy onion

Onion				
Model name	Avg accuracy	Max accuracy	Min accuracy	Fitness factor
On_Jan	84.44	99.09	57.18	1.20
ON_Feb	83.32	99.15	57.75	0.85
ON_Mar	93.34	99.73	84.77	1.55
ON_Apr	94.35	99.99	85.42	1.80
ON_May	93.42	99.96	74.46	1.75
ON_Jun	82.27	99.78	43.46	2.80
ON_Jul	57.82	99.04	13.44	1.95
ON_Aug	93.81	99.93	82.98	1.90
ON_Sep	75.78	99.86	36.47	1.90
ON_Oct	89.65	99.02	70.32	1.50
ON_Nov	88.17	99.49	71.92	1.50
ON_Dec	82.14	99.99	56.71	1.75
Average	84.88	99.59	61.24	1.70

6 Conclusion

The researchers constructed the model with fitness coefficient-α to predict future market price (Gujarat, India) in the form of generated knowledge and by this means, tried to connect the farmer community to the agriculture market. This 'generated knowledge' surely helps farmer community in their decision making of farm management activities such as (i) Take investment decisions at farm level

(ii) Determine optimum planting date, (iii) Evaluate weather risk and (iv) Determining best choice of cultivars to plant in their farm. Thus, this study helps rural people (specifically farmer community) and thereby in betterment of rural development.

References

1. Planning Commission Government of India: Twelfth Five Year Plan (2012–2017) Economic Sector Chapter 12 'Agriculture', vol. II. Sage Publication (2013)
2. Planning Commission Government of India: Twelfth Five Year Plan (2012–2017) Economic Sector Chapter 17 'Rural Development', Vol. II. Sage Publication (2013)
3. Ghadiyali, T., Lad, K., Patel, B.: Agriculture intelligence: an emerging tool for farmer community. In: Proceedings of Second International Conference on Emerging Application of Information Technology EAIT, vol. 2, pp. 313–316. IEEE, Kolkata (2011). doi:10.1108\9/EAIT 2011.36
4. Khater, E.S.G., Bahnasawy, A.H., Ali, S.A.: Mathematical model of compost pile temperature prediction. J. Environ. Anal. Toxicol. **4**(6) 1–7 (2014). doi:10.4172/2161-0525.1000242
5. Ramesh, D., Vishnu Vardhan, B.: Data mining techniques and applications to agricultural yield data. Int. J. Adv. Res. Comput. Commun. Eng. **2**(9), 3477–3480 (2013)
6. Nasrin Fathima, G., Geetha, R.: Agriculture crop pattern using data mining techniques. Int. J. Adv. Res. Comput. Sci. Softw. Eng. **4**(5), 781–786 (2014)
7. Brandt, J.A., Bessler, D.A.: Price forecasting and evaluation: an application in agriculture. J. Forecast. **2**(3), 237–248 (1983). doi:10.1002/for.3980020306
8. Haciyev, P.: Application of mathematical modeling methods on investment decisions in agro industrial company. Middle-East J. Sci. Res. **10**(2), 183–187 (IDOSI Publications, Moscow) (2011)
9. Ghamari, S., Rabbani, H., Khazaei, J.: Mathematical models for predicting the terminal velocity of Chickpea, Rice and Lentil. World Appl. Sci. J. **15**(11), 1557–1561 (IDOSI Publications, Tehran) (2011)
10. Allen, S.J., Schuster, E.W.: Controlling the risk for an agricultural harvest. Manuf. Serv. Oper. Manage. **6**(3), 225–236 (2004). doi: 10.1287/msom.1040.0035
11. Bai, P., Banks, H.T., Dediu, S., Govan, A.Y., Last, M., Lloyd, A.L., Nguyen, H.K., Olufsen, M.S., Rempala, G., Slenning, B.D.: Stochastic and deterministic models for agricultural production networks. Math. Biosci. Eng. **4**(3), 373–402 (2007)
12. Govardhana, R.G., Solmonrajupaul, K., Vishnu Sankarrao, D., Dayakar, G.: Seasonal variations and forecasting in wholesale prices of rice (paddy). Guntur District of Andhra Pradesh. Int. J. Dev. Res. **4**(11), 2418–2422 (2014)
13. Ticlavilca, A.M., Feuz, D.M., McKee, M.: Forecasting agricultural commodity prices using multivariate Bayesian machine learning regression. In: Proceedings of the NCCC-134 Conference on Applied Commodity Price Analysis, Forecasting, and Market Risk Management. St. Louis, MO. April 19–20 (2010). http://www.farmdoc.illinois.edu/nccc134
14. Hoffman, L.: Using futures prices to forecast the season-average U.S. Corn Price. In: Proceedings of the NCR-134 Conference on Applied Commodity Price Analysis, Forecasting, and Market Risk Management St. Louis, April 19–20 (2004)
15. Kaur, M., Gulati, H., Kundra, H.: Data mining in agriculture on crop price prediction: techniques and applications. Int. J. Comput. Appl. **99**(12), 1–3 (2014). IJCA: www.ijcaonline.org
16. Huang, W., Huang, Z., Matei, M., Wang, T.: Price volatility forecast for agricultural commodity futures: the role of high frequency data. Rom. J. Econ. Forecast. **4**, 83–103 (2012)
17. Raorane, A.A., Kulkarni, R.V.: Review-role of data mining in agriculture. Int. J. Comput. Sci. Inf. Technol. **4**(2), 270–272 (2013)

18. Yethiraj, N.G.: Applying data mining techniques in the field of agriculture and allied sciences. Int. J. Bus. Intell. 1(2), 72–76 (2012)
19. Agriculture Portal, Government of India. www.agmarknet.nic.in
20. Meteorological Department, Government of India. www.imd.gov.in
21. Weather and Meteorological data Portal. www.tutiimpo.in
22. Vishwanath, B.C, Madival, S.A., Madole, S.: Recognition of Fruits in Fruits Salad Based on Color and Texture Features. Int. J. Eng. Res. Technol. (IJERT) **1**(7), 1–6 (2012)

A Methodical Study on Behavior of Different Seeds Using an Iterative Technique with Evaluation of Cluster Validity

Karuna C. Gull and Akshata B. Angadi

Abstract Data analysis methods are vital for analyzing the rising colossal scale of high-dimensional data. Today, cluster analysis is a widely known technique applied and universally practised in many research areas. Among '*n*' different clustering techniques we briefly deal with centroid model, i.e., *K*-means, which is an iterative clustering technique. The recital of this algorithm is dependent on certain factors, which include the selection of initial centroid and the approach used in performing reckoning from each data point to different cluster centers. Initial pattern considered randomly by *K*-means algorithm often make the clustering results reach the local optima, i.e., choice of initial seed (pattern) greatly affects the ultimate clusters that results, in terms of inter and intra cluster distances and firmness. In this research paper author experimented the behaviors of different patterns with different distance metrics on *k*-means. Finally estimated validity check, i.e., cluster division ratios for every distance measure used and patterns considered. The experimental grades showed the maximum cluster parting and observed better cluster quality when chosen the initial seed as per the assumptions made, compared to patterns randomly picked. The pragmatic results were met when tried with different sample set and different *k* values. Further an addition of automatic detection of ideal initial pattern as mentioned by author statements leads to an additional trait to *k*-means.

Keywords Centroid · Cluster analysis · Initial cluster centres (ICC) · *K*-means

---

K.C. Gull (✉)
K.L.E. Institute of Technology, Hubli, India
e-mail: karunagull74@gmail.com

A.B. Angadi (✉)
MVJ College of Engineering, Bangalore, India
e-mail: akshata_angadi@yahoo.co.in

A.K. Saini et al. (eds.), *ICT Based Innovations*, Advances in Intelligent Systems and Computing 653, https://doi.org/10.1007/978-981-10-6602-3_7

1 Introduction

A good clustering algorithm supremely must produce groups/clusters with dissimilar nonoverlapping boundaries. Cluster analysis [1–3, 8] endeavor to pass through data swiftly to gain first order data by dividing data points into different groups (disjoint) such that data points fitting into same cluster are alike while data points fitting into different clusters are unlike. *K*-means method is one of the most accepted and competent clustering method. It is an iterative technique that is sensitive to initial starting conditions. The *K*-means clustering algorithm is about approximation of the mean values of a set of *k*-groups. In this algorithm, assignment of instances to clusters will be order-sensitive.

Working: For data point x_1 in ($x_i = x_1, x_2, x_3, x_4 \ldots x_n$) in cluster/Group G_i, if the centroid C_i is the nearest indication point, no tuning are made and the algorithm proceeds to the next data point. Nevertheless, if the centroid C_j of the group G_j is the indication point flanking to data point x_1 in ($x_i = x_1, x_2, x_3, x_4 \ldots x_n$), then x_1 is reassigned to group G_j, the centroids of the "trailing" group G_i (minus point x_1) and the "gaining" group G_j (plus point x_1) are recalculated, and the indication points C_i and C_j are moved to their new-fangled centroids. After each step, every new value of the *k* reference points acts as a centroid, or mean, hence the name "*k*-means" [4]. A graphical representation of clustering steps using the standard *k*-means algorithm is shown in Fig. 1.

There are a number of variants of the *k*-means algorithm. In this paper, we have taken a sample data set, tested it, and compared with many different seeds positively with the *k*-means algorithm with random restarts and used different distance measures. If a deprived choice is made prior in preferring an initial seed value, the algorithm may encounter an instance t_i that has no possible valid cluster [5].

Paper is organized as: Sect. 2 discusses about different distance measures used in paper. Section 3 details the methodology. In Sect. 4, how do we analyze *k*-means algorithm by taking the Posts as sample data is briefly explained. Lastly in, Sect. 5 Conclusion with future work is discussed.

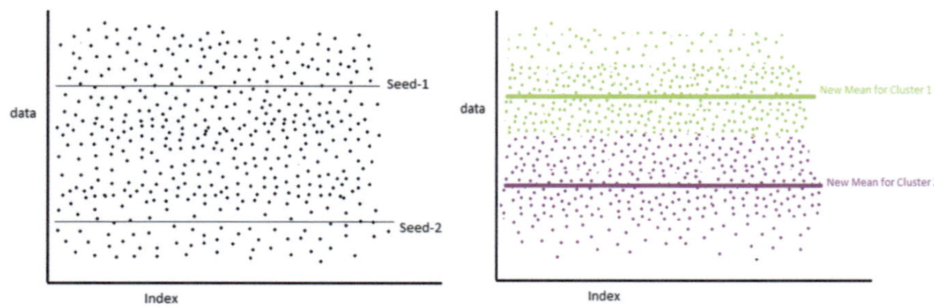

Fig. 1 *K*-means procedure [9]

2 Discussion

Clustering is the method of probing a collection of objects and grouping those objects into clusters based on some distance measure [6, 10]. The objective is to put objects in the same basket/cluster that are having small distance from other, while objects with larger distance from other are placed in different clusters [6, 10].

2.1 *Distance Measures*

Let the distance between two *x* and *y* (both vectors) be *D*(*x*, *y*). We now define a number of distance measures

Euclidean distance

The distance between data vector *x* and centroid *v* is computed as

$$D(x, v) = \sqrt{\sum (x_i - v_i)^2} \tag{1}$$

Manhattan distance

The distance between data vector *x* and centroid *v* is computed as

$$D(x, v) = \sum |(x_i - v_i)| \tag{2}$$

Chebyshev distance

This distance metric is based on the maximum attribute difference.

$$D(x, v) = \max |(x_i - v_i)| \tag{3}$$

Chi square distance

The distance between data vector *x* and centroid *v* is computed as

$$D(x, v) = \sqrt{\sum \frac{(x_i - v_i)^2}{(x_i + v_i)}} \tag{4}$$

2.2 *Theory of Standard* K-*Means Algorithm*

- Initial cluster seeds are chosen (at random).
- Suitable distance measure principles are applied and computed from each object of every cluster and each object is assigned to the nearer cluster as shown in Fig. 1.

- For each cluster, the new centroid is computed and each seed value is replaced by the respective cluster centroid.
- If Manhattan Distance is taken,
  - The Manhattan distance, i.e., shortest path an object could get to arrive at each cluster/group is computed, and the object is assigned to the cluster with the smallest distance. The cluster centroids are recalculated based on the new membership assignment.

Above three steps are repeated until no object moves clusters.

3 *K*-Means Method

K-means is a simplest and most popular classical clustering method. The method is called K-means because each K cluster is represented by the objects/the centroids within it. Also called as the centroid method as at each step the central point of each cluster is assumed and the remaining points are allocated to that cluster whose centroid is bordering to it. Once this part is completed, the centroids of a cluster are recalculated using *k*-means. The process of assigning points to every cluster is continual until there is no change in a huddle like stopping norm is met.

3.1 K-*Means Algorithm*

The *K*-means algorithm is described as follows:

Let $X = \{x_1, x_2, \ldots, x_n\}$ be the set of data points and $v = \{v_1, v_2, \ldots, v_c\}$ be the set of cluster centers.

1. Select the numbers of clusters (Consider '*k*').
2. Choose '*k*' seeds as the centroid of the clusters (ICC). The seeds have to be picked randomly unless user has some insight into the data.
   Assumption: We begin by considering the *k* Initial cluster centres as initial seeds for *k* clusters. Based on observations on real data, choose seeds that are possible to be in different clusters/groups. Each *k* seeds (ICC) selected initially, should belong to different clusters respectively.
   Example if $k = 3$ (clusters) and if assumed seeds are k_1, k_2 and k_3 then k_1 should belong to a Cluster that k_2 and k_3 does not belong. Similarly, k_2 should belong to a Cluster that k_1 and k_3 doesn't belong. Lastly, k_3 should belong to a Cluster that k_2 and k_1 doesn't belong.
3. Compute the different distances from each of the centroid.

4. Based on the distances computed in previous step, assign every object to the nearest points. Next, compute the new centroids of the clusters by computing its mean.

$$v_i = \frac{\sum x_i}{c_i} \tag{5}$$

5. Check if the stopping norm has been met. If yes go to step 6 else go to step 3. At this stage stop or rip the cluster or combine two clusters until ending criterion is met.

3.2 *Cluster Analysis*

Various cluster analysis approaches are available. Among them one possible approach is finding largest value of *E*/*I* ratio from all available results here '*E*' is between cluster variation and '*I*' is Average within cluster variation.

To find *I* and *E*:

$$I = \frac{\sum I_i}{k} \tag{6}$$

$$E = \frac{\sum_i \sum_j (\mu_i - \mu_j)^2}{k^2} \tag{7}$$

where

$$I_i = \frac{\sum (x_j - m_i)^2}{M_i} \tag{8}$$

and

$$m_i = \frac{\sum x_j}{M_i} \tag{9}$$

where

M_i	Number of objects in cluster C_i
k	Number of clusters
m_i	Mean of the cluster C_i
x_j	Number of attributes in each Object
$(\mu_i - \mu_j)$	Pairwise distances between the centroids of k clusters

To state if '*I*' is small leads to tight clusters and '*E*' is small with respect to I connotes the partition of clusters is difficult, which results in bad clustering. Thus good separation between clusters is possible if either '*E*' value is large or '*E*/*I*' ratio is largest.

4 Analysis of *K*-Means Algorithm

For the analysis we have taken 8 posts (Facebook) and will classify them into three different categories. The categories/groups chosen by us for Facebook Posts are Sports (Grp 1), Advertisement (Grp 2) and Education (Grp 3), i.e., $K = 3$. The specified K value will determine the number of clusters that are created.

To classify the posts into categories:

Let $M = \{m_1, m_2, m_3, m_4, \ldots m_n\}$ represents n messages (posts content) to be classified. Each of these messages, m_i, is defined by the 's' special words, i.e., $m_i = \{m_{i1}, m_{i2}, m_{i3} \ldots m_{is}\}$. Here, each mi is a s-dimensional vector representing the 's' special words which will be normalized using equation

$$\text{Occurrences} = \text{WC} / \text{TC} * 100 \tag{10}$$

where,

WC Word Count
TC Total no. of words in a message.

To select limited number of special features we set Threshold Matching Strength [7]. Thus, the final sample data is shown in Table 1. For $k = 3$.

4.1 *Analysis When* K = *3*

Step 1: Begin with considering the first three seeds as initial cluster centres (as pattern 1) from data set from Table 1.

Table 1 The sample data set taken from Facebook for analysis

Special words	Post-1	Post-2	Post-3	Post-4	Post-5	Post-6	Post-7	Post-8
Mobile	10	15	500	350	550	10	15	20
Cadbury	15	7	380	450	360	15	10	20
Books	50	60	70	60	10	540	450	390
Exams	60	80	50	40	20	380	430	410
Players	400	350	10	15	10	10	30	30
Court	500	250	15	20	15	15	40	20

Step 2: Using Eqs. (1 to 4) compute the distances between all special features of posts and assumed seeds for variety of distance measures (like Euclidean, Chebyshev, Chi square, etc.). The distance values for all objects (posts) and for all distance measures are tabulated in Table 2, from initial three seeds respectively.

Step 3: Based on distances each data (post) is assigned to nearest cluster. We obtain the final results once the stopping condition has met for pattern 1 for different distance measures. Table 3 shows the assignment of data to a particular cluster for pattern 1, after several iterations.

Step 4: Repeat the above steps (1 through 3) for other patterns (Pattern 2 through Pattern 7).

As per the observation and the requirement of author, the posts 1 and 2 must belong to cluster 1, posts 3, 4 and 5 must belong to cluster 2 and

Table 2 Initial cluster centres with distance values for different distance measures

Pattern 1					
Cluster centres (CC)					
Cluster centre—1		Cluster centre—2		Cluster centre—3	
10		15		500	
15		7		380	
50		60		70	
60		80		50	
400		350		10	
500		250		15	
Euclidean			Chebyshev		
From CC1	From CC2	From CC3	From CC1	From CC2	From CC3
0	256.1035	872.4391	0	250	490
256.1035	0	739.0393	250	0	485
872.4391	739.0393	0	490	485	0
827.0127	689.3468	166.2829	480	443	150
895.0698	766.6544	86.02325	540	535	60
854.298	700.9379	838.5255	490	480	490
803.3991	648.9291	813.6031	460	390	485
Chi Square			Manhattan		
From CC1	From CC2	From CC3	From CC1	From CC2	From CC3
0	9.712981	40.49727	0	343	1760
9.712981	0	36.78931	343	0	1473
40.49727	36.78931	0	1760	1473	0
39.15611	35.40649	5.997402	1670	1383	250
41.38241	38.04404	7.796065	1840	1573	160
38.30665	33.36268	37.72863	1685	1368	1655
36.21618	30.99199	37.51883	1610	1273	1660

Table 3 Belongingness of post/s to cluster/group (Grp) after several iterations, for different distances and for pattern 1

Post no.	Euclidean (E)	Chi Square (CS)	Chebyshev (C)	Manhattan (M)
Post 1	Grp1	Grp1	Grp1	Grp1
Post 2	Grp2	Grp2	Grp2	Grp2
Post 3	Grp3	Grp3	Grp3	Grp3
Post 4	Grp3	Grp3	Grp3	Grp3
Post 5	Grp3	Grp3	Grp3	Grp3
Post 6	Grp2	Grp2	Grp2	Grp2
Post 7	Grp2	Grp2	Grp2	Grp2
Post 8	Grp2	Grp2	Grp2	Grp2

posts 6, 7 and 8 must belong to cluster 3 for dataset considered in Table 1.

In this case we have considered 7 different patterns (1 through 7), among which the first 6 patterns (1, 2, 3) (2, 3, 4) (3, 4, 5) (4, 5, 6) (5, 6, 7) (6, 7, 8) do not meet the condition specified in the assumption (mentioned in Sect. 4) and if next pattern, i.e., pattern 7 (1,3,6) is chosen as combination of these ((1, 2) (3, 4, 5) (6, 7, 8)), then it is fulfilling the specified assumption (mentioned in Sect. 4).

Thus the final results, i.e., belongingness of each data to cluster after several iterations, for different distances and for all patterns are shown in the Tables 4 and 5. To talk more about the same if we choose initial seeds belonging to different clusters/groups (Assumption in Sect. 4), the result is as per the requirement of the author. And it is also better analyzed by using cluster analysis which is done in step 5.

Step 5: Calculate the E (Between Cluster Variations) and I (Average within cluster variation) values for all distances and patterns. E/I ratio shows the cluster separation. Higher the E value compared to I, will lead to good cluster division. Hence we pick the max value for E/I ratio that leads to final or expected results, i.e., placement (or finding the belongingness) of post/s to a particular cluster. Table 6 shows the ratio of E/I for all patterns and distances.

As per cluster analysis we observed that the E/I ratio is maximum for pattern 4 and pattern 7 (initial assumptions) which yield in a required result, i.e., the posts 1, 2 should belong to cluster-1, posts 3, 4, 5 should belong to cluster-2 and posts 6, 7, 8 should belong to cluster-3. When we see the outcome of the pattern 7 using distance measures like Euclidean, Chi Square, Chebyshev, and Manhattan shows the perfect, expected result with ratio 39.39 (Table 6) compared to other patterns. The pattern 4 which is the combination of posts 4, 5, 6 (where in 4, 5 are belonging to C2 and 6 to C3) shows the good result for Manhattan distance measure with ratio (37.33) that shows better cluster division compared to other distance measures. This shows the dependency of K-means with the initial seeds we chose. The point that we are highlighting here is the dependency factor of K-means and the comparison of

Table 4 Belongingness of post/s to cluster or group (Grp) after several iterations, for different distances and for 6 different patterns

Post no. ↓	Euclidean					Chi Square				
Patterns →	2	3	4	5	6	2	3	4	5	6
Post 1	Grp1	Grp2	Grp1	Grp3	Grp3	Grp1	Grp2	Grp3	Grp3	Grp3
Post 2	Grp1	Grp2	Grp1	Grp3	Grp3	Grp1	Grp2	Grp3	Grp3	Grp3
Post 3	Grp2	Grp1	Grp2	Grp1	Grp3	Grp2	Grp1	Grp1	Grp1	Grp3
Post 4	Grp3	Grp2	Grp2	Grp1	Grp3	Grp3	Grp2	Grp1	Grp1	Grp3
Post 5	Grp2	Grp3	Grp2	Grp1	Grp3	Grp2	Grp3	Grp2	Grp1	Grp3
Post 6	Grp1	Grp2	Grp3	Grp2	Grp1	Grp1	Grp2	Grp3	Grp2	Grp1
Post 7	Grp1	Grp2	Grp3	Grp2	Grp2	Grp1	Grp2	Grp3	Grp2	Grp2
Post 8	Grp1	Grp2	Grp3	Grp2	Grp3	Grp1	Grp2	Grp3	Grp2	Grp3
Post No. ↓	Chebyshev					Manhattan				
Patterns →	2	3	4	5	6	2	3	4	5	6
Post 1	Grp1	Grp2	Grp1	Grp3	Grp2	Grp1	Grp2	Grp1	Grp3	Grp3
Post 2	Grp1	Grp2	Grp1	Grp3	Grp3	Grp1	Grp2	Grp1	Grp3	Grp3
Post 3	Grp2	Grp1	Grp2	Grp1	Grp3	Grp2	Grp1	Grp2	Grp1	Grp3
Post 4	Grp2	Grp2	Grp2	Grp1	Grp3	Grp3	Grp2	Grp2	Grp1	Grp3
Post 5	Grp2	Grp3	Grp2	Grp1	Grp3	Grp2	Grp3	Grp2	Grp1	Grp3
Post 6	Grp3	Grp2	Grp3	Grp2	Grp1	Grp1	Grp2	Grp3	Grp2	Grp1
Post 7	Grp1	Grp2	Grp3	Grp2	Grp2	Grp1	Grp2	Grp3	Grp2	Grp2
Post 8	Grp1	Grp2	Grp3	Grp2	Grp3	Grp1	Grp2	Grp3	Grp2	Grp3

Table 5 Belongingness of post/s to cluster (Grp) after several iterations, for different distances and for pattern 7

	Euclidean	Chi Square	Chebyshev	Manhattan
Post 1	Grp1	Grp1	Grp1	Grp1
Post 2	Grp1	Grp1	Grp1	Grp1
Post 3	Grp2	Grp2	Grp2	Grp2
Post 4	Grp2	Grp2	Grp2	Grp2
Post 5	Grp2	Grp2	Grp2	Grp2
Post 6	Grp3	Grp3	Grp3	Grp3
Post 7	Grp3	Grp3	Grp3	Grp3
Post 8	Grp3	Grp3	Grp3	Grp3

different distance measures. All distance measures proved better for both $k = 3$ and $k = 4$ with maximum *E*/*I* ratio (good cluster separation) iff the initial seeds are assumed properly.

Thus, we conclude that, *K*-means technique works perfectly if initial seeds are chosen properly. As it is very hard to set the initial seeds, need to try with all combinations of seeds and the output, i.e., placement of object/s to cluster, desires to be checked with respect to authors requirement and max *E*/*I* value. But if we

Table 6 The ratio of *E/I* for all patterns and distances

	Euclidean							Chi Square						
Patterns →	1	2	3	4	5	6	7	1	2	3	4	5	6	7
Ratio = *E/I*	12.36	5.13	3.13	3.84	2.28	2.55	39.39	12.36	5.13	3.13	5.19	2.28	2.55	39.39
	Chebyshev							Manhattan						
Patterns →	1	2	3	4	5	6	7	1	2	3	4	5	6	7
Ratio = *E/I*	12.36	6.58	3.13	3.84	2.28	1.71	39.39	12.36	5.13	3.13	37.33	2.28	2.55	39.39

choose seeds belonging to different clusters (Assumption in Sect. 4), and try with all supporting combinations of the same, and there is maximum possibility to get the best result (placement of object/s to cluster) with max *E*/*I* value, which also meets the authors' requirement. Thus if these results are fulfilling the needs of authors/users, then one can say that it is a better method to do the clustering for the kind of data samples they choose else give a try for other clustering methods like FCM [12], hierarchical clustering, etc., to come up with better results. To resolve this, a better option is to add intelligence to find the right seeds initially or opt for a different clustering methods like—Fuzzy techniques.

5 Conclusion

Evaluation in cluster analysis is an appealing problem as it produces different clusters (may not be meaningful) when used diverse clustering techniques. Hence, judging which is better is left to the user. User should analyze and recognize the result, which makes sense. In this paper, we detailed the procedure of centroid model, i.e., *K*-means. Variety of distance calculation methods can be used to compute distances between centroids and attribute values of objects to place those objects into one of the clusters using *K*-means method. In our case, with *K*-means algorithm we used distance calculation methods like Euclidean, Chebyshev distance, Manhattan distance, and Chi square distances to know their performance with sample set.

K-means algorithm can be contemplated as a gradient plunge method, which begins at starting cluster centroids, and iteratively updates the centroid values and retains same result at one point. In *K*-means Clustering technique, the shortcoming facet is finding the initial seeds or the pattern or centroids. The paper detailed about the analysis of finding the cluster separation depending on the different distance metrics and initial seeds using standard *k*-means algorithm on sample set. The experimental results showed varying cluster division for every different combination of patterns. But it worked perfectly for the patterns where we chose seeds belonging to different clusters, i.e., seeds selected initially should belong to different clusters (Pattern 7). We used the word perfect as it is matching the authors observed and expected result as specified in analysis section and its cluster validity ratio (for all distance measures) is higher, i.e., 39.39. For pattern 4 the results given by Manhattan, Euclidean, and Chebyshev is right but the *E*/*I* ratio (37.33, 3.84, 3.84, respectively) is lesser compared to pattern 7, whereas Chi square resulted in bad cluster separation.

Thus the analysis result state—"*K*-means clustering technique is dependent on the Initial Cluster centres a user chooses." The cluster validity ratio (*E*/*I*) reveals whether the division of clusters was better or not, i.e., Higher the ratio, higher is the possibility of good cluster separation. As a Future study, if intelligence is added initially in choosing a better initial seeds then it gives an excellent result with higher *E*/*I* ratio and good cluster division.

References

1. Duda, R.O., Hart, P.E., Stork, D.G.: Pattern Classification. Wiley, New York (2001)
2. Jain, A.K., Dubes, R.C.: Algorithms for Clustering Data. Prentice Hall, Englewood Cliffs, NJ (1988)
3. Hastie, T., Tibshirani, R., Friedman, J.: The elements of statistical learning—Data mining, inference, and prediction. Springer Series in Statistics, Springer-Verlag, New York, 2001. xvi + 533 pp. ISBN: 0-387-95284-5
4. Vance, F.: Clustering and the Continuous *k*-Means Algorithm. Los Alamos Science Number 22 1994
5. Wagstaf, K., Cardie, C., Rogers, S., Schroedl, S.: Constrained K-means clustering with background knowledge. In: Proceedings of the Eighteenth International Conference on Machine Learning, pp. 577–584 (2001)
6. Leskovec, J., Rajaraman, A., Ullman, J.D.: Mining of massive datasets. Course dealt by Stanford University. http://infolab.stanford.edu/~ullman/mmds/ch7.pdf
7. Gull, K.C., Angadi, A.B.: Systematic Approach using Fuzzy Clustering Technique towards betterment of the marketing proficiency—Facebook. In: Third National Conference on Advanced Technologies in Electrical and Electronic Systems (ATEES-14), pp. 323–332. Elsevier Publications (2014)
8. Reza, M.S., Ruhi, S.: Study of multivariate data clustering based on k-means and independent component analysis. Am. J. Theor. Appl. Stat. **4** (5), 317–321 (2015). doi:10.11648/j.ajtas.20150405.11
9. Psych993–Lectureon Clustering and Classification. http://jtemplin.coe.uga.edu/files/clustering/psyc993_09.pdf
10. Tarun Galyani blog Its all about Data. http://datachurn.blogspot.in/2013/05/clustering-tips.html, May 24, 2013

BharataNatyam Dance Classification with Rough Set Tools

Sangeeta Jadhav and Jyoti Pawar

Abstract BharataNatyam (BN) Choreography is known to be an intuitive domain. We have attempted to aid the choreographer for this creative domain of choreography for pure dance movements (used for aesthetics) called Nritta with our ArtToSMart (System Modeled art) system. Various automated techniques have been used for Western dance notations, its animation as well as for choreographic skills; but we have not been able to find enough on BN. We have used rough set tools to build a classifier for the dance poses, since manual methods are subjective and time-consuming. Thirty attributes have been used to define the human body. We obtained eight reducts from among these 30, and also decision rules which frame the grammar of a BN pose. The results are promising and have higher accuracy for about 500 instances. A comparative study has been done with two different tools namely WEKA (Waikato Environment for Knowledge Analysis) 3.6.11 and RSES (Rough Set Exploration System) 2.2.2 which has given 87.42% and about 76.8% classifier accuracy, respectively.

Keywords BharataNatyam · Grammar · Classification · Dance choreography · Dimension reduction · Rough set theory

1 Introduction

Indian Classical Dance (ICD), especially BharataNatyam (BN) has been a subject of fascination both in the east as well as the west. The beginning of this style can be traced back to developments in the medieval period, roughly dating from 1300 AD to 1800 AD [1]. The ancient scriptures of dance namely Natyashastra (NS) and

S. Jadhav (✉)
S. S. Dempo College of Commerce and Economics, Altinho, Goa, India
e-mail: dcst.sangeeta@unigoa.ac.in

J. Pawar
Department of Computer Science and Technology, Goa University, Taleigão, India
e-mail: jdp@unigoa.ac.in

A.K. Saini et al. (eds.), *ICT Based Innovations*, Advances in Intelligent Systems and Computing 653, https://doi.org/10.1007/978-981-10-6602-3_8

Abhinayadarpana (AD) help us in revealing the intricacies involved in this classical art form. A pure dance movement, termed as Nritta[1] in BN, involves the coordinated movement of the major limbs of the body namely; head, hands, waist, and legs. A simple combination of all the possibilities of the major limbs shows more than 5 lakh possibilities for a single beat [2]. Few of these are impractical, un-aesthetic, and also it is impossible for a human being to remember and perform all the possible moves. Thus a computerized tool will aid in the choreographic process to generate unpracticed moves but within the classical framework. We have used the following main steps for automation in BN choreography:

Modeling of the Human body: The first step was to model the human body [3]. We have used 30 attribute dance position (DP) vector to portray the position of a dancer's major limbs like head, right hand, left hand, waist, right leg, and left leg.

Choreography: The outputs obtained from the single beat choreography module [4] were used to generate multi-beat dance sequences [5]. The constraints were as follows: hand mudra[2] filter, leg filter, and fitness value filter. Adavus[3] in BN are the coordinated limb movements for Nritta. Studying the Adavu patterns revealed us how to use the above-mentioned filters to get the best choreographic pattern for multiple beats. We have studied all the techniques and patterns of Adavus which are considered to be ideal Nritta movements and developed unique but unpracticed choreography. These were evaluated by several domain experts.

Stick Figure Generation: The problem of visualizing the 30 attribute DP vector was challenging and hence we used the 2D stick figure model [6].

Dimension Reduction using Rough Set: Several dance experts who evaluated our results from the ArtToSMart system had ambiguous answers. To avoid further conflicts, we took an average of their ratings for the same dance pose and prepared a tagged database. These were tagged as OK, Good, or Bad as per each expert opinion and the Excellent data was picked from the existing repertoire of BN called as Adavus. Further we wanted to automate the task of classification of each dance pose and hence used RSES, a rough set tool. The system gave us reducts from the 30 attributes and several decision rules were obtained [7]. A comparative study was carried out with another tool WEKA for classification of the dance poses.

The current paper is organized as follows: The literature reviews for related work have been put forth in the next section followed by the data preparation method for BN dance. The problem statement has been specified in Sect. 4 followed by experimental results. The final section is on conclusion and future work.

[1]Nritta: rhythmic dance movements in BharataNatyam. It showcases geometric patterns and very complex rhythmic variations. http://www.natyakriya.com/2013/01/nritta-nritya-natya.html.

[2]mudra(s): A gesture done with Hand or Fingers for the Indian Classical Dance.

[3]Adavu(s): A basic step in BN used for pure dance movement.

2 Related Work

ICDs have been the fascinating point for many researchers in various areas like animation [8], heritage preservation [9], e-learning [10], notating [11], and mobile applications [12]. The use of multiple Hidden Markov Models to obtain an action classifier has been an inspiration from our work for the M.Tech. thesis of an IIT student [13]. Although several researchers have attempted to choreograph [14–16] animate [17–19], and notate [20] Western dances, we have not encountered any research work so far in the field of choreography for BN.

Feature selection through rough set based methods has been explained in [21, 22]. To the best of our knowledge, rough set tools have not been applied for the dance choreography domain.

3 BN Dance Data Preparation for RSES

Codification and Attributes: We have used 30 attributes to define a human pose which include various body parts namely the head, hands, waist, and legs. The codification of these attributes is done on the basis of their Sanskrit names as obtained from the NS and AD, which are the rule books of ICD. The first two attributes are for head (Shirobheda, Orientation), next eight attributes depict the right hand (mudra, *X*, *Y*, *Z* axis, elbow, wrist, palm positions, and shoulder jerk) followed by left hand, waist has two attributes (twist and bend) whereas right leg and left leg has six attributes (ankle, knee position, *X*, *Y*, *Z* axis), respectively. These 30 attributes are depicted as follows: [SB, ORI], [HAR, HRX, HRY, HRZ, HRE, HRW, HRP, HRJ], [HAL, HLX, HLY, HLZ, HLE, HLW, HLP, HLJ], [WT, WB], [AR, KR, LRX, LRZ, LRY], [AL, KL, LLX, LLY, and LLZ].

Instances and Decision Class: Initially we experimented with 224 objects which were tagged by dance experts. Out of these 130 instances were tagged as "Excellent" since they are taught by the teachers to a beginner dance student and are termed as data for pure dance movements only (Nritta). The remaining were tagged by domain experts. A Questionnaire Form was given to the teachers for evaluating the dance poses generated by our ArtToSMart system. It had "Good", "Bad", "OK", "Not Acceptable" as options. Later we omitted Not Acceptable from our decision class, since none of the dance experts had rated any of our system generated pose as Not Acceptable [23]. For the current work, we have used these 224 tagged instances and built a classifier with up to 501 instances. The Decision Class had four attributes namely OK, Bad, Good, and Excellent.

4 Problem Statement

In the linguistic sense, there would be no grammar rules, which would define whether a dance pose is valid or invalid. However, we can statistically formulate some classification rules that would classify a dance pose as Bad, OK, or Good. Thus for each such rule we will have to associate various parameters such as support, confidence, false positive, accuracy, etc. We are working on classifying individual dance poses and then based on these classification obtained we shall classify '*n*' beat dance sequences.

We have designed 30 attribute DP vector for depicting a dance pose. Thus our aim is to reduce these attributes to a minimal subset. We also need to retain high accuracy while determining a feature selection method in the field of BN. This shall help us in data analysis and determination of the most important attributes from among these 30 DP vectors.

The ArtToSMart system generates several enumerated dance poses every time. Manually tagging these by experts have shown that the task is very time consuming and also very subjective to every person. Thus to automate the task of classifying these dance sequences and also for the best feature selection, we have used Rough Set Exploration Tool (RSES 2.2.2) and WEKA 3.6.11.

5 Experimental Results

224 trained instances (validated by domain experts) were given as an input to RSES and reducts were obtained. Using cross-validation method, the system had generated 72.7% accuracy for the trained data [7].

Since we noticed that the earlier 224 trained instances had more data belonging to class 4 (termed as Excellent) and thus there was a tendency of the rules getting only the Class 4. We therefore had to increase the number of trained instances so that there was a proportionate distribution of all the classes (Fig. 1).

WEKA
Using WEKA, we trained the system up to 501 instances. In the earlier 224 trained instances we had 130 instances tagged as Excellent from Adavus data and 94 instances tagged as OK, Good, Bad based on an average obtained from the expert ranking. The classifier accuracy increased to 87.42% with these 500 instances as compared to 66.96% for 224 trained instances as seen in Table 1.

RSES
Using RSES, we got 76.8% accuracy for the 501 trained instances as compared to the earlier data which had generated 72.7% accuracy for the 224 trained data using cross-validation method, refer Table 1. Ten reducts were generated by RSES, refer Fig. 2, out of which only two of the reducts had seven attributes whereas eight of them had eight attributes. Thus we can consider that the final reducts obtained from the system were eight.

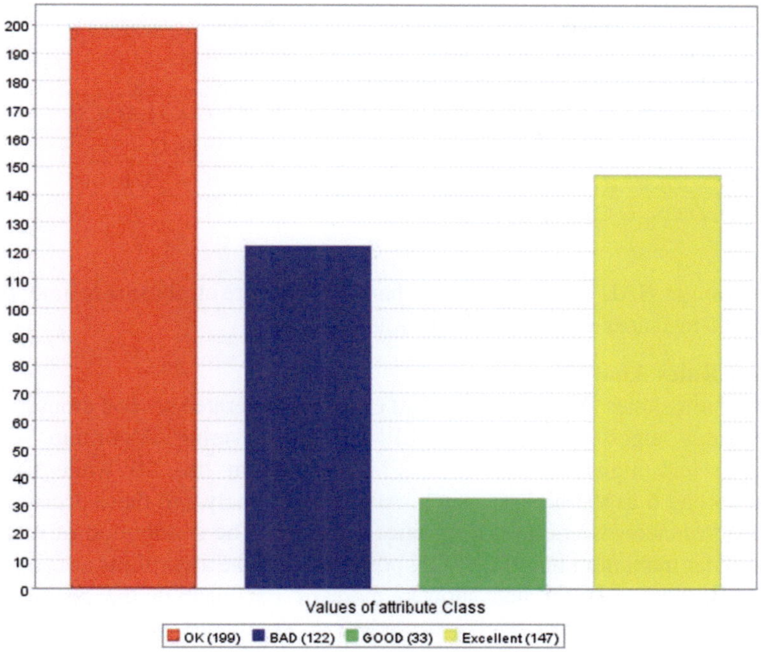

Fig. 1 501 trained instances obtained from the 224 expert tagged data

Table 1 Classifier accuracy comparison

	RSES		WEKA	
Instances	224	501	224	501
Classifier accuracy (%)	72.7	76.8	66.9	87.42

Reduct set: reducts from 500 instances

(1-10)	Size	Pos.Reg.	SC	Reducts
1	7	1	1	{ SB, HAR, HRW, HAL, HLY, AR, AL }
2	8	1	1	{ SB, HAR, HRZ, HRW, HAL, HLY, KR, LLX }
3	8	1	1	{ SB, HAR, HRX, HAL, HLP, AR, KR, LLX }
4	8	1	1	{ SB, ORI, HAR, HAL, HLY, HLW, AR, LLY }
5	7	1	1	{ SB, HRW, HAL, HLY, HLE, AR, LLY }
6	8	1	1	{ SB, HRX, HRZ, HAL, HLY, HLW, AR, LLX }
7	8	1	1	{ SB, HRX, HRW, HAL, HLY, AR, KR, LLX }
8	8	1	1	{ ORI, HAR, HAL, HLY, HLW, WB, AR, AL }
9	8	1	1	{ ORI, HAR, HRE, HAL, HLY, HLE, AR, AL }
10	8	1	1	{ ORI, HAR, HAL, HLW, HLP, AR, KR, LLY }

Fig. 2 Eight attributes obtained as reducts

Table 2 Statistics for rule set using various methods

Algorithm max. support min. support no. of rules generated				Rules distribution
LEM2	20	1	136	E = 25, O = 62, G = 11, B = 38
Genetic	50	1	5114	E = 1549, O = 1939, G = 376, B = 1259
Covering	6	1	191	E = 34, O = 82, G = 15, B = 60

E Excellent, *O* OK, *G* Good, *B* Bad

Also we got HAL (mudra of the left hand) as the core attribute this time. For the earlier 224 instances no core attribute was obtained.

Decision Rules Analysis
Four decision values OK, bad, good, and excellent generated several thousand rules with various algorithms: Genetic, LEM2 and Covering Algorithm. Genetic Algorithm had highest match 50, LEM2 Algorithm had 20 while Covering Algorithm had 6 as the highest match. After further analyzing these decision rules with 500 instances, we noticed a significant change. The distribution of rules now was showing maximum number of OK poses with all the algorithms as can be seen in Table 2. Earlier RSES had showed more of Excellent in the decision rules generated (Refer Sect. 5).

6 Conclusion and Future Work

A rough set tool, RSES was used for 224 trained instances. These were evaluated to find the best possible features and decision-making system for the classes. This helped in automating the process of dance pose classification. The system was trained further to get more accuracy and later tested with the enumerated data. The reducts obtained were used for the purpose of comparison with another tool WEKA and we got promising results here too. Analysis of the decision rules from these trained instances helped us to classify unknown dance poses automatically. Various statistics from RSES namely: support, confidence, false positive, accuracy, etc., are identified and being exploited for defining the grammar for BN dance. Apriori algorithm is also being experimented to generate appropriate association rules between the attributes. Finally domain experts can also validate all these results and hence we can get a robust choreographic tool for BN.

References

1. Sridhar, V.: Survival of Indian Classical Dance Down the ages. June (1996)
2. Jadhav, S., Sasikumar, M.: A computational model for BharataNatyam choreography. Int. J. Comput. Sci Inf. Secur. **8**(7), 231–233, Oct (2010)

3. Jadhav, S., Joshi, M., Pawar, J.: Modeling BharataNatyam dance steps: art to SMart. In: CUBE, pp. 320–325. Pune, Maharashtra, India (2012)
4. Jadhav, S., Joshi, M., Pawar, J.: Art to SMart: an evolutionary computational model for BharataNatyam choreography. In: 12th International Conference on Hybrid Information Systems. Pune, Maharashtra, India (2012)
5. Jadhav, S., Joshi, M., Pawar, J.: Art to SMart: automation for BharataNatyam choreography. In: 19th International Conference on Management of Data. Ahmedabad (2013)
6. Jadhav, S., Aras, A., Joshi, M., Pawar, J.: An automated stick figure generation for BharataNatyam dance visualization. In: International Conference on Interdisciplinary Advances in Applied Computing. Amritapuri, Coimbatore, India (2014)
7. Jadhav, S., Joshi, M., Pawar, J.: Towards automation and classification of BharataNatyam dance sequences. In: Int. J. Comput. Sci. Appl. Technomathematics Res. Found. **11**(2), 93–104 (2014)
8. Pattanaik, S.N.: A stylised model for animating BharataNatyam, an Indian Classical Dance form. In: Computers in art, design and animation, pp. 264–273. New York, Springer (1989)
9. Mallik, A., Chaudhury, S., Ghosh, H.: Nrityakosha: preserving the intangible heritage of Indian Classical Dance. ACM J. Comput. Cult. Heritage, **4**(3), 11:1–11:25, Dec (2011)
10. Hariharan, D., Acharya, T., Mitra, S.: Recognizing hand gestures of a dancer. In: International conference on Pattern Recognition and Machine Intelligence, pp. 186–192. Springer, Berlin Heidelberg (2011)
11. Karpen, A.: Labanotation for Indian Dance, in Particular BharataNatyam. In 11th European Conference. Amsterdam (1990)
12. Majumdar, R., Dinesan, P.: Framework for teaching BharataNatyam through digital medium. In: IEEE Fourth International Conference on Technology for Education (T4E), Hyderabad, India (2012)
13. Sharma, A.: Recognising BharataNatyam dance sequences using RGB-D data. I.I.T, Kanpur, India (2013)
14. Ebenreuter, N.: Transference of dance knowledge through interface design. In: Extended Abstracts Proceedings of the 2006 Conference on Human Factors. Montreal, Canada (2006)
15. Hsieh, C.M., Luciani, A.: Generating dance verbs and assisting computer choreography. In: Proceedings of the 13th ACM International Conference on Multimedia, Singapore, Nov 6–11 (2005)
16. Kar, R., Konar, A., Chakraborty, A.: Dance composition using microsoft kinect. In: Transactions on Computational Science XXV, Springer, pp. 20–34 (2015)
17. Nahrstedt, K., Bajcsy, R., Wymore, L., Sheppard, R., Mezur, K.: Computational model of human creativity in dance choreography. In: Creative Intelligent Systems, Papers from the 2008 AAAI Spring Symposium, pp. 53–60. California, USA (2008)
18. DeLahunta, S.: The future of choreographic practice: the choreographic language agent. Feb (2009)
19. Calvert, T., Wilke, L., Ryman, R., Fox, I.: Applications of computers to dance. In IEEE Xplore. Apr (2005)
20. Rett, J., Dias, J., Ahuactzin, J.-M.: Bayesian reasoning for Laban movement analysis used in human-machine interaction. Int. J. Reasoning-based Intel. Sys. 13–35 (2009)
21. Swiniarski, R.W., Skowron, A.: Rough set methods in feature selection and recognition. In: Pattern recognition letters, pp. 833–849 (2003)
22. Bazan, J.G., Nguyen, H.S., Nguyen, S.H., Synak, P., Wrblewski, J.: Rough set algorithms in classification problem. In Rough set methods and applications. Springer, pp. 49–88 (2000)
23. Jadhav, S., Joshi, M., Pawar, J.: Art to SMart: an automated BharataNatyam dance choreography. Appl. Artif. Intel.: Int. J. **29**(2), 148–163. Taylor and Francis (2015)

Effective and Efficient Digital Advertisement Algorithms

Vishal Assija, Anupam Baliyan and Vishal Jain

Abstract In today's world when everyone wants the best deal when it comes to online shopping, it becomes tedious job for companies to target their audience with appropriate content and advertising media. This paper talks about such issues faced by various companies nowadays and tries to resolve those issues by implementing some algorithms or by introducing a new platform to showcase advertisements online. In this we also talk about efficient way of tracking users online and analyze their data to showcase content based on their preferences. Our focus is also on the usage of "AdBlock" plug-in, that is, how to stop it or prevent it to block advertisements online. At last we try to deliver advertisements in such a way that they should take as less as bandwidth they can and increase the page load time.

Keywords PPC · ISP · PPV · CPM · CPC · CPA · RTB · DMP · CPL · ASCII · Bots · DSP · ROI

1 Introduction

The rise of the Internet has substantially changed the advertising industry's business models as well as their client basis. One of the most striking innovation is how to sell advertisement space online. Advertisements in the form of banners, are sold through the traditional model, i.e., cost per impression, but also with methods based on a visitor taking some specifically defined action in response to an advertisement. This research examines the way in which advertiser can showcase the ads online as well as pricing strategy of web publishers operating in a market where they are

V. Assija (✉) · A. Baliyan · V. Jain
Bharati Vidyapeeth's Institute of Computer Application and Management, Delhi, India
e-mail: vishalassija007@gmail.com

A. Baliyan
e-mail: anupam_hod1976@yahoo.co.in

V. Jain
e-mail: vishaljain_usit@yahoo.in

A.K. Saini et al. (eds.), *ICT Based Innovations*, Advances in Intelligent Systems and Computing 653, https://doi.org/10.1007/978-981-10-6602-3_9

unable to influence the price of the advertisement (i.e., whether cost per impression is better or cost per action) and effectiveness and efficiency of algorithms that are used to display advertisements online.

Traditionally, advertisers sell online space to marketing companies where they can showcase their banner and in lieu of that need to pay for the space provided. But now this has been revolutionized by the technologies like "Google AdWords & AdSense", where one can monitor how much revenue he/she is able to make via particular Ad campaign.

So, to increase user interaction we need to understand his behavior and activities well so that the content which user likes will be shown at appropriate place, so that the chances of user engagement increases. Advertiser will be charged only if user clicks or sees the advertisements, i.e., pay-per-click (PPC) or pay-per-view (PPV). After the introduction of these type models, advertiser starts investing more in online space, since no user means no charges, while on the other hand, more users engagement means more revenue for the advertiser as well as successful advertisement campaign.

While some of the users already have "ADBLOCK" plug-in for their browser, it is difficult to show advertisement on webpages. But as you know, Internet is full of ideas to overcome this problem we have "Pop-Ups" (they also can be blocked but up to some extent). My idea is to integrate advertisements deeply into webpage that none of the plug-in can block it and on the other side the bandwidth used to deliver the ads will be minimized.

Main focus is on how to increase the revenue of online advertiser by displaying relevant ads to particular user and make user engagement, so that the hit ratio will increase tremendously. But to keep in mind that the users' concerns like privacy and what data related to particular user, where, and how to be used still needs some answers from advertisers. In long run, it seems to be great model to increase user engagement by keeping all the user interests safe.

2 Literature Review

If you look at late 90s when Internet was at its boom, someone introduced an innovative idea of displaying ads, i.e., "Banner" and the inventor is HotWired. At that time, browsers are not smart enough to block pop-ups that are displaying ads, so at that time this model is very successful as it gives good ROI to advertisers. But it has its own consequences like slow page loads because of loading the extra images to be shown as advertisement and script related to these ads (it seems relevant because displaying ads over a dial up connection takes too long to load page and relatively slow speeds that are provided by ISPs). It was just the beginning of new era of online marketing and advertisement which no one could ever imagined of and its fueled the portal war, whether reaching to customer in a traditional way is better or online where advertiser will only be charged if customer clicks on their banner ads. At that time banner ads powered the web first explosion of advertisements and generating revenue for websites owners. So, in association with HotWired, Wired Magazine introduced a new search engine called HotBot which basically targets audience with specific ads and it was generating more than $25 million a year in revenue. That amazed everyone and attracts the attention of venture capitalist to start investing in it and secure the elusive "early mover advantage".

And then four students from Harvard invented a search engine called "Google" (currently). Once it gained popularity, Google started finding ways to generate revenue by its millions of users to visit its site and search anything they wanted to know or explore. They came with an idea of showing relevant ads to users corresponding to their search queries or frequent behavior on website. By implementing such model, Google earned a lot of money (to be specific billions of dollars from its search advertising revenues). It is time for Google to evolve as a big online advertisement aggregator in the United States of America to surpass the Yahoo.

First, most concern of every advertisers is to reach brand safe environments and well-lit areas also known as publishers without porn. And publishers are expecting some good amount of revenue from advertisers that do not harm their brand either. So, it is the start of complication, because there are so many places for advertisers to buy and for publisher to sell.

To avoid such complications and to increase the process transparency in publishing of advertisements online, there is a need of intermediary companies which will overlook at all these things and provide high-quality ads to be published via publishers. The main business of these is to provide online space via buy/sell model and distinguish themselves as either DSPs (Demand Side Platform) or SSPs (Seller Side Platform).

To understand how diverse the advertisement space online is, please take a look at picture given below

If you try to figure out how these companies are making money on the Internet, then you will land in a pool which seems to be filled with water but when you jump on it, you realize that it is empty. Since generating revenue by displaying ads online is not as easy as it sounds. It requires a lot of skills and energy to get some good amount of revenue. You can take example of any big money making business say, legal, financial, medical, gaming, etc., that are going to be focused on online advertisements rather than traditional marketing. You can say that this is the Internet version of Wall Street jive. So make sure you understand all these acronyms listed below before exploring online advertisements world.

CPM (Cost Per Impression): It simply means cost to be incurred for an online advertisement per thousand views. Here M stand for 1,000 (i.e., Roman Numeral).

CPC (Cost Per Click): It is also known as PPC and it is called so because, as user hits on the banner ads only then advertiser will be charged. This is an online ads model used to direct traffic to websites (traffic refers to human being), where advertiser will be charged by publisher only if the user clicks on the ad displayed in banner. Some of the biggest players in this category are: Google AdWords, Yahoo! Search Marketing, and Microsoft adCenter. All of these operate under the model which is called "Bid-Based". That simply means advertiser is on safe side, they can pay only if user clicks on their ads, else they will not be charged and it creates an effective way of displaying ads online.

CPL (Cost Per Lead): In this model of online advertisement, advertiser pays only for a sign-up from (which is explicit) a consumer interested in ads offer. If you compare this to the aforementioned models, advertisers gain advantage of legitimate qualified sign-ups by only interested users and making Cost-Per-Lead models the holy grail of the Internet advertising ROI hierarchy.

CPA (Cost Per Action): As the name suggests, in this model advertiser needs to pay only if specific action is performed by the user. (Actions like: click on the banner, filling of sign-up form, user engagement on a page for more than 5 min, purchase from some specific merchant, etc.). Action-based model attracts advertiser who prefers direct response from users (as in brand marketing it is not required).

RTB (Real Time Bidding): In this type of model advertiser and publisher meet in real time in the online marketplace for the buying and selling of ad impressions.

DMP (Data Management Platform): It forms the backbone required by the advertiser to collect user data for different advertisement operations. It works a cross bridge between servers that collect user data and the servers that displays advertisements to the users (it is used to build customer relationship management tools).

3 Related Work

To understand the user behavior online we have used "Google Analytics Tool" on a website. After using it for a while we have seen that online marketing and advertisements companies can track users rigorously and they have all the information about a particular user that they need to display ads. Some of the screenshots of Analytics are shown below

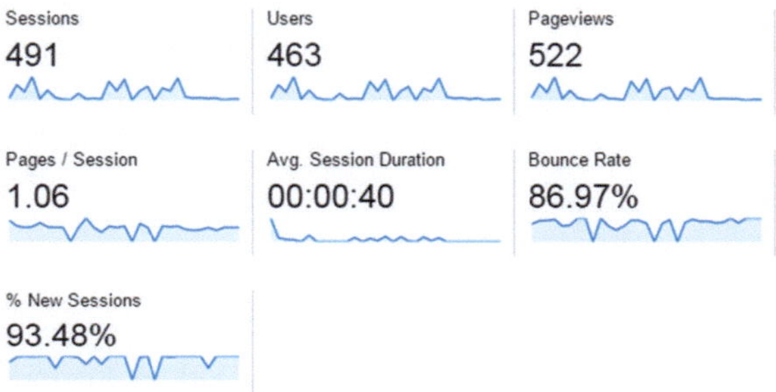

The above picture shows the total number of sessions, users, and page views. It also tracks the session duration to know which page has high user engagement. Using Google Analytics we can also track which browser is used by users to open our website, so that we can optimize our website according to the users' browsers usage history. And if you have enabled specific publisher account conversion ratio within Google analytics tool, then it will also shows how much revenue is generated by the browsers listed below.

Browser	Acquisition			Behavior	
	Sessions	% New Sessions	New Users	Bounce Rate	Pages / Session
	491 % of Total: 100.00% (491)	97.35% Avg for View: 93.48% (4.14%)	478 % of Total: 104.14% (459)	86.97% Avg for View: 86.97% (0.00%)	1.06 Avg for View: 1.06 (0.00%)
1. Chrome	450 (91.65%)	98.00%	441 (92.26%)	90.22%	1.10
2. (not set)	26 (5.30%)	100.00%	26 (5.44%)	30.77%	0.31
3. Firefox	11 (2.24%)	100.00%	11 (2.30%)	100.00%	1.00
4. Safari	4 (0.81%)	0.00%	0 (0.00%)	50.00%	2.00

We can also track users' geographic locations to cater the needs of users better. For example, currently our website is in English language only but we are having huge traffic from Asia (India and China), in that case we can translate out website content to the specific languages to make user happy and in the end it will increase the traffic as well as user engagement.

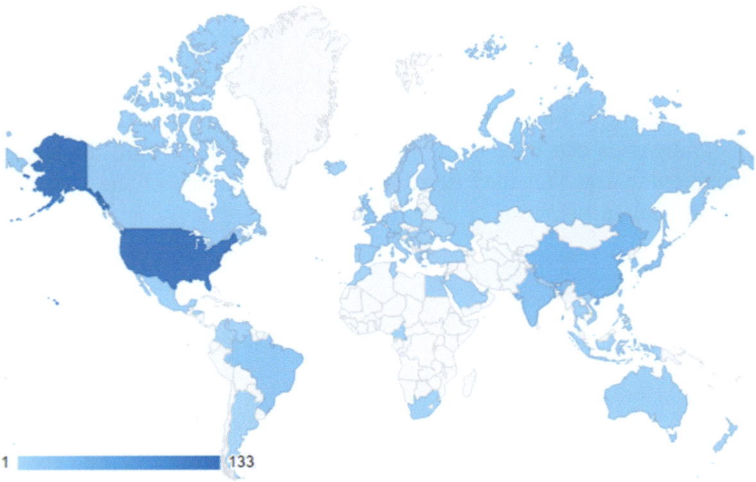

It shows user engagement on the basis of sessions. Dark blue color shows high user engagement from that particular geographic location and as the color fades, the user will also decrease. It also shows all this information in tabular form. (shown below)

Country	Acquisition		
	Sessions	% New Sessions	New Users
	491 % of Total: 100.00% (491)	97.35% Avg for View: 93.48% (4.14%)	478 % of Total: 104.14% (459)
1. (not set)	133 (27.09%)	100.00%	133 (27.82%)
2. United States	122 (24.85%)	100.00%	122 (25.52%)
3. China	35 (7.13%)	100.00%	35 (7.32%)
4. India	25 (5.09%)	64.00%	16 (3.35%)
5. Japan	19 (3.87%)	100.00%	19 (3.97%)
6. Brazil	17 (3.46%)	100.00%	17 (3.56%)
7. South Korea	13 (2.65%)	100.00%	13 (2.72%)
8. United Kingdom	12 (2.44%)	100.00%	12 (2.51%)
9. Germany	8 (1.63%)	100.00%	8 (1.67%)
10. Russia	8 (1.63%)	50.00%	4 (0.84%)

4 Problem Statement

While the number of Internet users increases tremendously, digital advertisement industry also grows rapidly. Although there are some catches to it. Like users are more aware of their data privacy that advertisements companies try to steal, and they know how prevent these things. For marketing companies, users data is very important to plan their campaigns and future advertisements plans but private mode and dynamic IP are those problems that need to be taken care of; otherwise it is impossible to find out whether the request made by a new user or previous user.

Privacy is one concern and other than that is how to target specified media in the form of advertisement that would make a great impact on user. Simply means how to design such ads that would cater the need of user as well as marketing company. Interactive ads require "Flash Player" which most of the users do not have, so it means your ad will not displayed.

Once you have user data the new problem that arises is how to use it to deliver ads that will suite user preferences. User data may contain—DOB, Name, Gender, Site visited, Product Watched Page Liked, etc. So how to use this data as an effective tool to deliver ads that will suite user requirements is a big question and need to be answered.

AdBlock plug-in is other problem for digital advertisers, it will simply block advertisements on a webpage and remove any third-party link associated to it. Pop-Up blockers are already there to stop showing ads in a pop-up window. So all these things need to be resolved, otherwise all these things will be making a great loss to online advertisers.

5 Proposed Solution

When we talk about online advertisements, one thing came to our mind—these advertisements were for users who are online. So all of the advertisement algorithms treat user preferences at utmost priority and deliver only those ads through which user will be benefited. Another thing that needs to be transparent is user data and user privacy. Which data advertiser is using while users visiting any site and how they are going to utilize it must be clear, so that user can understand the privacy policy easily.

Advertisements take a lot of bandwidth while transferring and page load takes longer time than usual non-advertisement page. This needs to be resolved; otherwise ads will take a large chunk of useful bandwidth that will use to deliver some important information. This can be done via compressing ad banners and send them a string of characters. At user side it will be processed by browsers and will show some animated ASCII characters Ads which attracts users and helps in increasing revenue for online advertisers.

Another thing to keep in mind is how to differentiate legitimate user and fake bots online. Fake bots are built to submit spurious information that will affect the advertisers' policy of showcasing a particular ad to a group of users. So identifying fake bots is also a big challenge and need to be implemented as soon as possible.

Advertisements blockers and plug-in which facilitates the blocking of advertisements are need to be taken care of. Every year, billion dollars revenue loss are done by these kinds of plug-ins. In respect of user terms, they serve the purpose of blocking advertisements, but on the other hand, users will miss the great opportunity of finding a great deal online according to their preferences.

6 Conclusion/Future Research

A lot of research is required to deal with all these problems related to online advertisements; and proposed solutions may provide temporary solution to the problems stated, but in long term there may arise new problems which would affect the revenue of advertisement companies in many ways. To overcome this, we need a standard policy for online advertisements which should be followed by all the companies and all the Internet users must be aware of how their data is being utilized by these companies to display advertisements.

References

Website References

1. http://goo.gl/HGPD21—10 Aug. 2015 14:00 Hrs
2. https://goo.gl/IiKcv5—10 Aug. 2015 15:30 Hrs
3. http://goo.gl/PsvJyj—9 Aug. 2015 11:00 Hrs
4. http://goo.gl/dPQ5W2—11 Aug. 2015 13:30 Hrs
5. http://goo.gl/JQKDu—12 Aug. 2015 10:15 Hrs
6. http://goo.gl/Lxm16Z—11 Aug. 2015 20:00 Hrs
7. Allenby, G.M., Leone, R.P., Jen, L.: A dynamic model of purchase timing with application to direct marketing. J. American Stat. Assoc. **94**(446), 365–374 (1999)
8. Black, J.: Online advertising: it's just the beginning. BusinessWeek Online (July 12). Available at http://www.businessweek.com/technology/content/jul2001/tc20010712_790.htm (2001). Accessed 10 Oct 2005
9. Pechmann, C., Stewart, D.W.: Advertising repetition: a critical review of wearin and wearout. In: Leigh, J.H., Martin, Jr., C.R. (eds.) Current Issues and Research in Advertising, vol. 11 (1–2), pp. 285–329. Ross School of Business, University of Michigan, Ann Arbor (1988)
10. Hoffman, D., Novak, T.: When Exposure-Based Advertising Stops Making Sense (and What CDNOW Did About It). Working paper, Owen Graduate School of Management, Vanderbilt University (2000)

11. Zufryden, F.: A model for relating advertising media exposures to purchase incidence behavior patterns. Manage. Sci. **33**(10), 1253–1266 (1987)
12. Sherman, L., Deighton, J.: Banner advertising: measuring effectiveness and optimizing placement. J. Interact. Mark. **15**(2), 60–64 (2001)
13. Song, Y.: Proof That On-line Advertising Works. Atlas Institute. Available at https://iab.net/resources/pdf/OnlineAdvertisingWorks.pdf. Accessed 12 Oct 2005 (2001)
14. Dahlen, M.: Banner advertisements through a new lens. J. Advertising Res. **41**(4), 23–30 (2001)

Automation of Patient Information in Healthcare System

Rishav Shaw and K. Govinda

Abstract Patients spend much time in hospitals for non-functional activities such as registering themselves, depositing money, and then collecting bands/tokens from the reception. Manual efforts such as taking the readings and recording it again take time. Hospitals lose some patients (i.e., money) and patients lose their time due to inefficient system. The Automated Healthcare System consists of an NFC system, sensors (temperature, pressure, and ECG), ZigBee communication module and a terminal (i.e., a display device). The system will authenticate the patient through his/her NFC card. It will take the patient's readings (temperature, pressure, and beat rate) by sensors and display it in the LCD screen. The communication module, ZigBee, will then transfer the readings wirelessly to the doctor's PC.

1 Introduction

A patient spends much time standing in the queue for the non-functional activities such as registration, money deposition, and token/band collection in hospitals. The patient then has to wait in the queue for cash deposition with his hospital card if he applies for reappointment when he comes for the next time. These activities take up a lot of patient's time and also hospital's time. An efficient system could have reduced the time taken for such activities resulting in the more number of patients checked up per day. More number of patients per day also increases per day profit.

Tests such as temperature, pressure, and beat rate readings involve a bit more of manual efforts. A compounder has to watch all the readings and note them down the patient's medical report. The human intervention again reduces the speed of operation. This can be sorted out with the help of microcontroller, sensors and

R. Shaw (✉) · K. Govinda
School of Computing Science and Engineering, VIT University, Vellore 632014, India
e-mail: rishav2105@gmail.com

K. Govinda
e-mail: kgovinda@vit.ac.in

A.K. Saini et al. (eds.), *ICT Based Innovations*, Advances in Intelligent Systems and Computing 653, https://doi.org/10.1007/978-981-10-6602-3_10

actuators (LCD and display). The readings taken up by the sensors will be shown on the microcontroller's LCD and on the doctor's PC.

Some famous hospitals in India such as Christian Medical College (CMC) has the online system but they only let customers to fill registration form and get their hospital number. They do not let customers pay money through net banking but provide the facility of credit card payment through telephone.

The payment is done by calling to the bank and then providing them card details and OTP (one-time password) through telephone. It usually takes about 20 min of time in telephone to exchange information and make the payment. Overall from form filling to making payment it takes roughly 45 min of time which is quite significant in today's world. Even after paying up the money, a patient has to come to the reception and stand in line to get the band/token.

Automated healthcare system provides a solution to save the patient's and hospital's time with the help of NFC tags. The NFC tag would get registered when the patient comes for the first time. NFC tag will provide all the necessary information to the hospital required for registration just at the swipe of the card. The patient will get authenticated by the swipe of this card over an NFC reader in the laboratory and then the readings will be taken and shown into the LCD. These small introduction of technologies in healthcare can speed up the processes many fold benefitting both patient and hospital.

2 Architecture

The goal of the Automated Healthcare system is to identify a patient, read the health parameters from his body, and transfer these readings wirelessly to the doctor's PC. In order to do this it should authenticate the patient by his NFC card, take the readings (temperature, pressure and beat rate) by the appropriate sensors, display it in the LCD and then transfer the readings to the doctor's PC with the help of ZigBee wireless communication module

The project has been divided

- Reading the NFC id from the NFC tag by swiping it over the NFC reader which continuously keeps sending RF waves.
- Authenticating the user by checking his NFC id with all available id's in the microcontroller.
- Reading the health parameters (temperature, pressure, and beat rate) through the use of appropriate sensors.
- Sending the information obtained by sensors to the distant terminal using wireless communication module, i.e., ZigBee.

The Automated Healthcare System is designed with patients and receptionists in mind. The end customer would not necessarily have a technical background, simplicity would be one priority. There will be different patients every time so the following functions will have to be performed:

- The NFC id will be read by the reader and sent to the microcontroller.
- Sensors will read the readings and the microcontroller will store those readings for that particular id, i.e., patient.
- The readings will be shown on the LCD screen and transmitted to the doctor's PC through ZigBee. ϒ At last the received readings will be shown in the PC (Fig. 1).

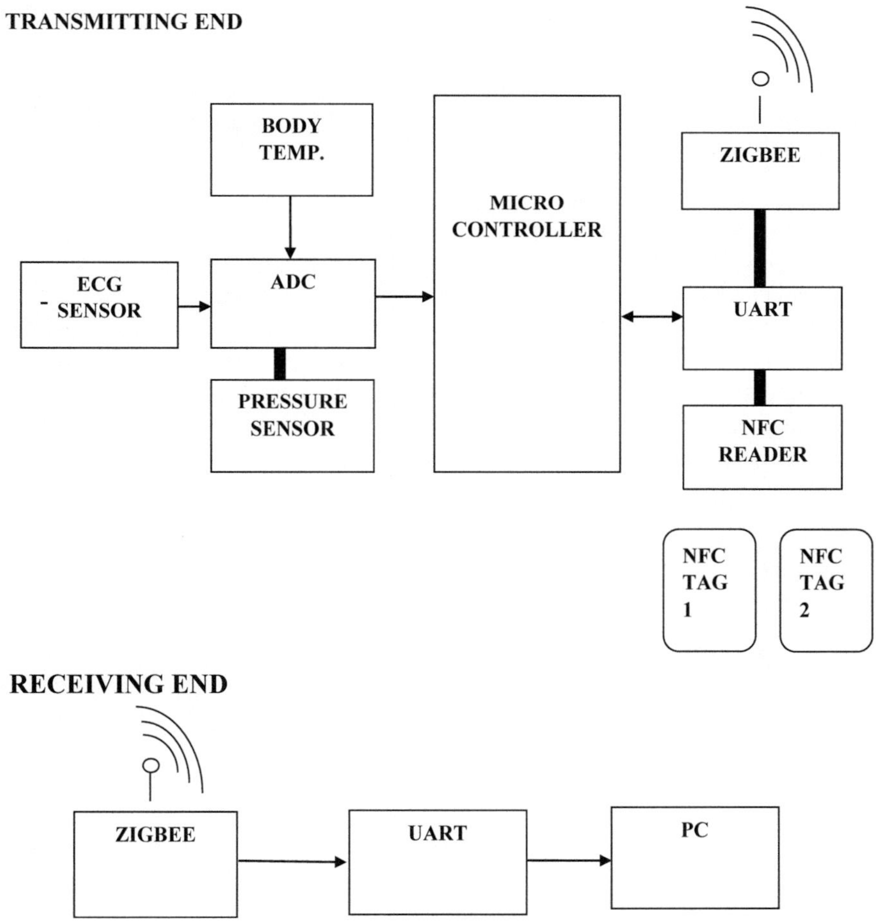

Fig. 1 Block diagram of healthcare system

3 Technical Specifications

While the goals and strategies have been stated, the specifications of the components will be determined as they are identified for their applicability in the project. The overall system will meet the specifications stated in Table 1.

4 Design Approach and Details

4.1 Near Field Communication (NFC)

The Automated Healthcare System uses an NFC system to identify the patient from the patients registered in the microcontroller. The NFC card acts as an identity for a patient. The utilities of this sensor are

- Identification—The patient will be identified by his NFC card. The NFC card will contain his information and when brought in-front of the NFC reader, the microcontroller will search the particular patient, i.e., his NFC id, in it and will provide access to the patient for further process if authentic.
- Information storage—NFC card stores the information of a patient. It can also be used to store the reports of a patient if needed. NFC can also be used to store the valuable information such as credit card/debit card numbers as from [1]. It can also be used to make payments as it is one of the most secure forms of the payment because of its working at very close proximity.

All the readings taken will be noted down under a particular id, i.e., NFC id which is unique for every card (Fig. 2).

The readings sent to the doctor's PC will be for a particular NFC id, i.e., for a particular patient.

Temperature, Pressure and ECG sensor:
Temperature sensor used for the process is LM35. This sensor is generally used as a prototype to show the temperature. The temperature sensor is meant to take the readings and send it to the ADC which will further be sent to the microcontroller. The readings taken are shown as volts on the LCD. With **LM35**, temperature can be

Table 1 Technical specifications

	Values
Operational specification	
Power supply	0–5 V
Performance specifications	
Temperature accuracy [7]	±0.5 C
Pressure accuracy [7]	97.5%
Failures per trial	1%
ZigBee data rate [1]	20–250 kbit/sec

Fig. 2 NFC reader

measured more accurately than with a thermistor. It also possess low self-heating and does not cause more than 0.1 °C temperature rise in still air [2].

The MP3V5050 series piezoresistive transducer is a state-of-the-art monolithic silicon pressure sensor designed for a wide range of applications, but particularly those employing a microcontroller or microprocessor with A/D inputs. It has a thermoplastic surface mounted package. This sensor has been used in the project to showcase the pressure readings. Any pressure put up into it will be recorded by it as from [3] (Fig. 3).

It determines the electrocardio graph using the difference in electrical energy measured from the electrodes placed in both arms taking voltage from right leg as

Fig. 3 Temperature sensor (LM 35)

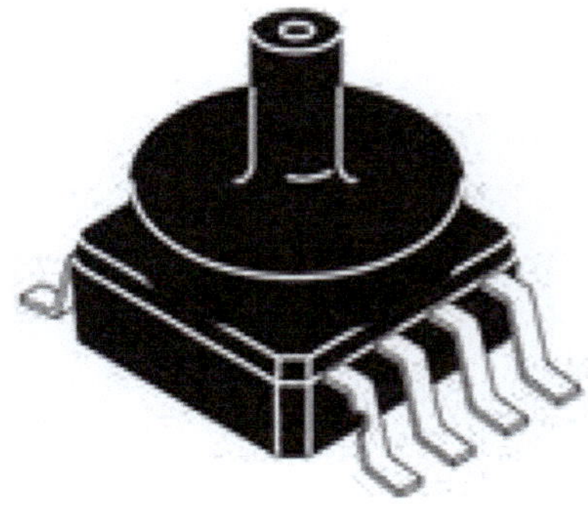

Fig. 4 Presser sensor

reference. The electrocardiogram (ECG) has grown to be one of the most commonly used medical tests in modern medicine. Its utility in the diagnosis of a myriad of cardiac pathologies has been invaluable to clinicians for decades. It uses three Ag/AgCl electrodes. The difference in voltage between two arms is taken and the right leg is taken as the reference [4] (Fig. 4).

Wireless Transmission:
There will be a wireless ZigBee transmitter mounted on the Automatic Healthcare System's main microcontroller. The transmitter will transmit the sensor values from the microcontroller's end. There will be a wireless receiver interfaced to a computer that will receive the health parameters read from the body by the sensors and will display it in the hyper-terminal (Fig. 5).

Voltage Control
An ADC has been used to convert the analog readings taken by the sensors into the digital form (as microcontroller understands only binary language) and sends them to the microcontroller for further process.

The project is connected to the power supply which works at 220 V AC which is further reduced to 12 V by a step down transformer. This 12 V is further converted to 5 V dc for the functioning of sensors (Fig. 6).

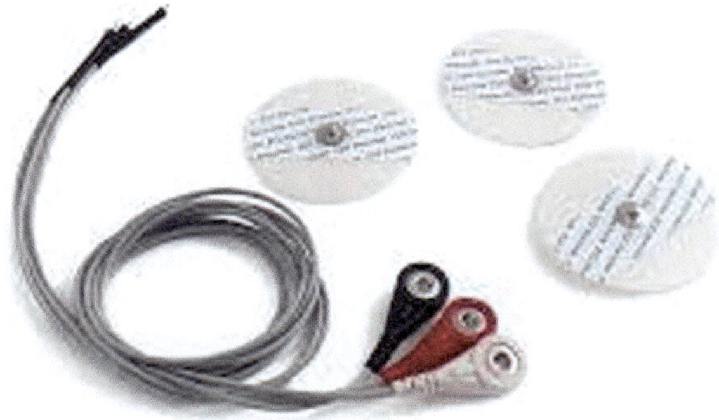

Fig. 5 ECG sensor

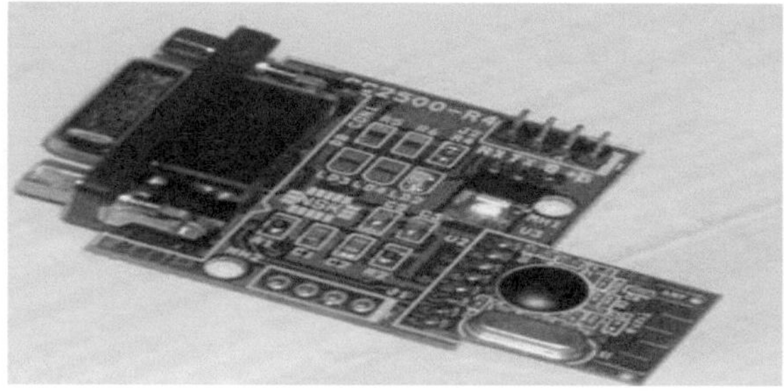

Fig. 6 Zigbee

5 Proposed Methodology

The most important specification that the Automated Healthcare System has to meet is to register a patient in the microcontroller connected to the NFC reader. The microcontroller should be able to identify the patient when the NFC card is brought over the NFC reader and the reader sends the information to the microcontroller. First bring a registered card over the reader and check how good the system works, then bring an unregistered card over the reader and then check its working, i.e., the microcontroller should not accept the card.

After the card authentication, the microcontroller will show the sensor readings. The sensors will be checked for the different conditions and the readings will be analysed (Fig. 7).

The last demonstration will be of the communication module, i.e., ZigBee used in the Automated Healthcare system. The readings will be transmitted from the transmitting part, i.e., the microcontroller part to the receiving part, i.e., the PC side. The blinking green LED in the ZigBee transmitter will show that the transmission is taking place (Figs. 8, 9, 10 and 11).

6 Result

Patients have to wait for a long time in the queue to get registered in a hospital, i.e., fill form, deposit money, and get hospital bands for check-up.NFC card can be used to store information of the patients. It (NFC) can be used as an identity, i.e., NFC id, for a particular patient. Readings of small tests such as temperature, pressure, beat rate, etc., are usually taken separately and noted down into the report. This requires a lot of manual effort which can be reduced by the use of sensors, microcontroller, and some display device.

To overcome these problems, a prototype has been designed using NFC technology which will simplify the whole process of health care for the patients.

The prototype has been successfully developed and tested. Designed system is validating a user by his/her NFC tag. After authentication, outputs of the sensors are

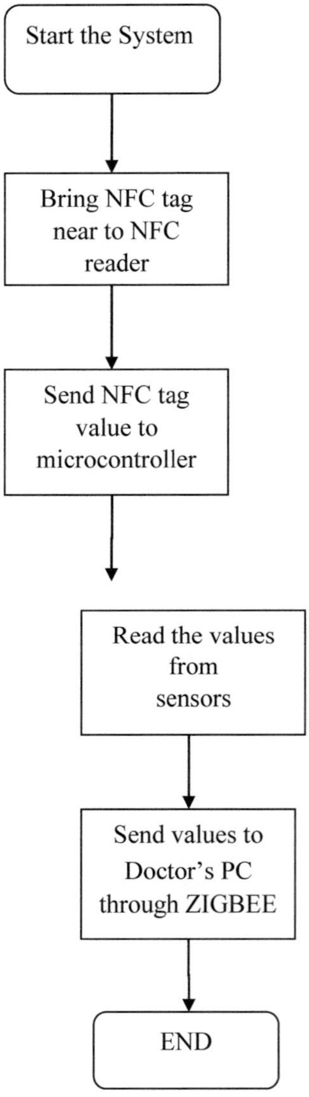

Fig. 7 Control flow diagram of automated healthcare system

being stored in microcontroller. The measured parameters are being shown onto LCD and being sent to distant terminal through ZIGBEE. The prototype successfully withstood long operational hours. No component failure was observed. The software of the prototype also worked very efficiently.

Nowadays some smart phones are NFC-enabled, so in future we can store the information in our mobile phones itself. This smart phone will keep the NFC card secure as they will be password protected. NFC as a promising short range wireless

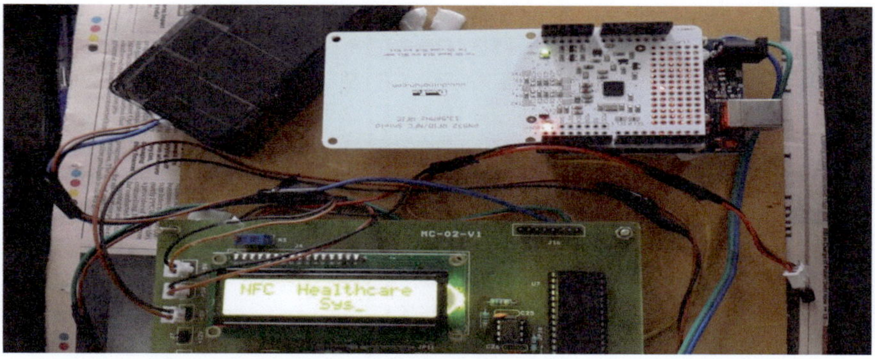

Fig. 8 Demonstration of NFC reader

Fig. 9 Demonstration of sensors

Fig. 10 Demonstration of ZigBee

communication technology facilitates mobile phone usage of billions of people throughout the world that offers diverse services ranging from payment and loyalty applications to access keys for offices and houses [5].

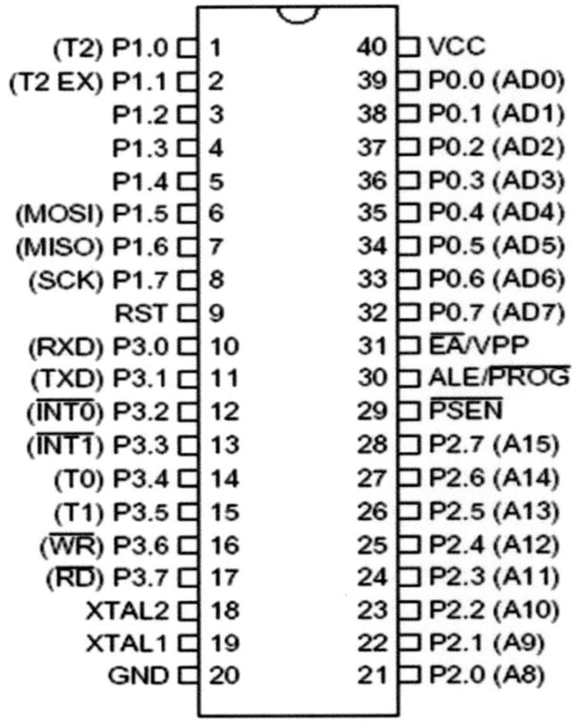

Fig. 11 PIN configuration of AT89S52 [6]

7 Summary

Patients have to wait for a long time in the queue to get registered in a hospital, i.e., fill form, deposit money and get hospital bands for check-up. Money deposition again involves credit and debit cards which might get misplaced while juggling through wallet in the process. NFC (Near Field Communication) card can store information of the patient along with the credit card details or Google Wallet Information for payment. It (NFC) will quickly register a patient by taking his information at a go and act as an identity, NFC id, for a particular patient. Small tests such as temperature, pressure, beat rate, etc., are usually taken separately. The readings are carefully tracked and then noted down into the report. This requires a lot of manual effort which can be reduced by the use of sensors, microcontroller, and some display device. The temperature, pressure, and beat rate can be measured by the respective sensors, sent to the microcontroller and then display it in the LCD screen. The medical report can further be sent to the doctor's PC wirelessly from laboratory (where the readings are being taken). This will help doctor to access patient's report while check-up without requiring patient to carry them with him. The medical report will be sent from the microcontroller to the doctor's PC through ZigBee communication module.

References

1. NFC Forum Technical Specification. http://members.nfc-forum.org/specs/spec_list/
2. http://www.facstaff.bucknell.edu/mastascu/eLessonsHTML/Sensors/TempLM35.html
3. http://www.vernier.com/products/sensors/bps-bta/ (pressure)
4. http://www.saelig.com/MFR0113/SEE012.htm (ECG)
5. Coskun, V., Ozdenizci, B., Ok, K.: A Survey on Near Field Communication (NFC) Technology. J. Wirel. Pers. Commun.: Int. J. **71**, 2259–2294 (2013)
6. Alldatasheet. http://www.alldatasheet.com
7. http://ubisensetech.com/index.php?page=shop.product_details&flypage=flypage.tpl&product_id=166&category_id=37&option=com_virtuemart&Itemid=1
8. Divyanshikha, Gupta, D., Mittal,T., Arora, U.: NFC based secure mobile healthcare system
9. NFC Wikipedia. http://en.wikipedia.org/wiki/Near_field_communication
10. ZIGBEE Wikipedia. http://en.wikipedia.org/wiki/ZigBee
11. Zigbee Wireless Standard Technology –Digi International http://www.digi.com/technology/rf-articles/wireless-zigbee

Recommendation for Selecting Smart Village in India Through Opinion Mining Using Big Data Analytics

Brojo Kishore Mishra, Abhaya Kumar Sahoo and Rachita Misra

Abstract In India, most of the people are staying in below poverty line. Nowadays, village people are also most inadequate with mobile phones. To develop this village as smart, we emphasize on different factors like agriculture, employment, nutrition security, environment, natural resource utilization, and conservation, etc. We select smart village based on collection of different opinion from village people in terms of forms, questionnaire, views and surveys, etc. Opinion mining extracts useful knowledge about village from ample of opinions. This mining process gives us a right direction for creating smart village. In this paper, we are trying to create digitalized village which is basically an application of Information and Communications Technology to define the major function of Government in order to bring about Small, Moral, Accurate, Reliable and Transparent. To get accurate response, we use big data analytic concept after mining opinions using map reduce approach.

Keywords Big data analytics · Map reduce · Opinion mining · Smart village

1 Introduction

In India, maximum people are living below the poverty line. To develop the culture of that village, we should look positively into some parameters such as education, agriculture, environment, road development, natural resource management like water preservation and soil preservation, etc. In spite of India is a developing country, nowadays most of the villagers are using digital devices like mobile phone, computer, calculator, and television, etc. But 40% people in India are extremely poor. Now India Government has declared that 2500 number of smart villages will be developed by 2019 under the Saansad Adarsh Gram Yojana (SAGY) scheme. The main objective of building smart village is to bring benefits to villagers in all

B.K. Mishra (✉) · A.K. Sahoo · R. Misra
Department of Information Technology, C.V. Raman College of Engineering, Mahura, India
e-mail: brojokishoremishra@gmail.com

A.K. Saini et al. (eds.), *ICT Based Innovations*, Advances in Intelligent Systems and Computing 653, https://doi.org/10.1007/978-981-10-6602-3_11

fields and to solve migration problem. Develop smart village concept can increase entrepreneurship in the villages. This program can focus on proper resource utilization, empowered local self-governance, access to assured basic amenities, and responsible individual and community behavior to build a vibrant and happy society in villages [1]. In this paper, we are going to choose the village to be developed as SMART. So, at first, we have to do survey in different villages through opinion mining. Opinion mining is the field of study which helps to collect people's opinion, sentiments, appraisals, attitudes, evaluations, and emotions toward a particular topic by doing questionnaire, forms, discussion panel, etc. [2]. Here opinion mining can be applied on villagers to get their opinions and views. These views collected from people are normally made in an unstructured way. To make it as structured, we need map reduce approach which can be applied on a huge amount of collected data called big data from people in Hadoop environment. Hadoop is an open-source framework that allows to store and process big data in a distributed environment using map reduce approach. Hadoop Distributed File System (HDFS) is associated with Hadoop which is purely based on Java. Hadoop automatically indexes the file, breaks the file into blocks and stores it in different nodes. After storing in HDFS, it requires map reduce approach for processing [3]. Our job is to easily get valuable people's opinion which corresponds to basic criteria of smart village development for making digital India.

This paper summarizes the followings: Sect. 2 describes related work which tells about basic criteria for establishing smart village and basic concepts of opinion mining and big data analytics. Section 3 tells how to select a village as smart village by opinion mining using map reduce approach. Section 4 explains implementation of proposed method to satisfy our problem statement. Section 5 shows experimental analysis and results. Section 6 presents conclusion.

2 Related Work

In the paper, we present the history of upcoming smart village program. The creation of smart villages in Africa and India has enabled by the study of sustainable energy provision undertaken by Malaysian Commonwealth Studies Center (MCSC), the European Academies Science Advisory. In smart villages, energy access takes a main role as a catalyst for development, enabling education and local business opportunities, improving health and welfare, and enhancing democratic engagement [4]. Electrical Research and Development Association (ERDA) is a cooperative research institution created by the Indian Electrical Industry and Utilities with the support of Governments of India and Gujarat which has taken main role as the smart village product development partner [5]. To choose a better village making to be smart village, we need an important survey on various numbers of villages. Then Government will take care of that village to build as smart village. The CGIAR Research Program on Climate Change, Agriculture and

Food Security (CCAFS) is working with communities to develop "Climate-Smart Villages" [6].

The paper [7] "An Study and Analysis of Opinion Mining and Sentiment analysis" written by Nayak and Fernandes summarizes as: Opinion mining is a process of extracting the people's views about a particular topic under discussion, organization and product. These views are collected through social media and review sites like blogs, comments, tweets or reviews. Using opinion mining concept, we can extract valuable factors which will be taken as main criteria of smart village from villagers.

The paper [3] reveals about concepts of map reduce approach of Hadoop which is a very efficient distributed processing framework. This framework is very efficient tool for storing and processing unstructured, semistructured and structured data. This concept can be used after collecting views from people through opinion mining.

3 Smart Village Selection by Opinion Mining Using Map Reduce

A. Concept of Smart Village

Now ICT has taken major role as drivers of growth in infrastructure, human capital, human education, social capital, relational capital, and environmental factors. Indian Government is going to developing the smart village in recent future. We have to give more focus on different factors which will be most necessary for establishing smart villages in India.

Main criteria of the smart village should be satisfied among villagers. These are written as below.

- Criteria-1: Improve Employment (C1)
- Criteria-2: Focus on Agriculture (C2)
- Criteria-3: Nutrition Security (C3)
- Criteria-4: Water Conservation (C4)
- Criteria-5: Soil Conservation (C5)
- Criteria-6: Removal of Environment Pollution (C6)
- Criteria-7: Facility of Education (C7)
- Criteria-8: Transport Facility (C8)
- Criteria-9: Woman Welfare (C9)
- Criteria-10: Wi-Fi Facility (C10)
- Criteria-11: Electric Facility (C11)
- Criteria-12: Facility of Computer Education (C12)
- Criteria-13: Mobile Communication (C13)

These above factors are required to setup smart village. First of all, we have to collect feedback from villagers on above factors.

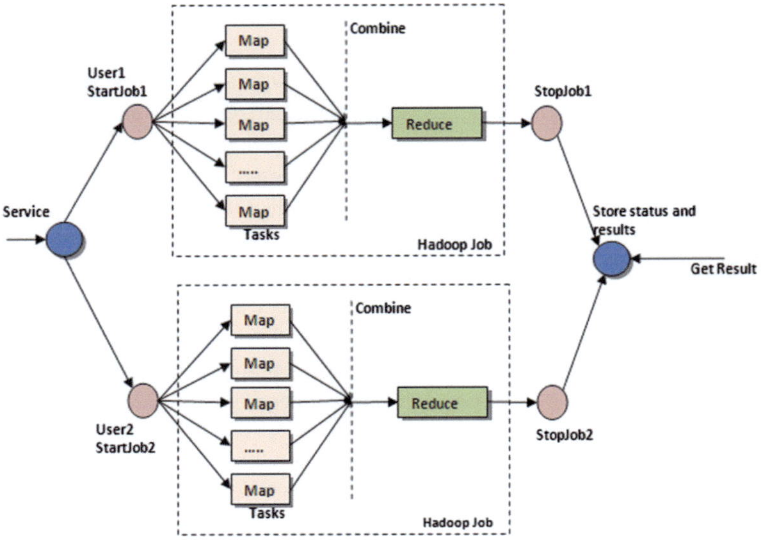

Fig. 1 Map reduce approach

B. **Opinion Mining Concept Using Map Reduce Approach**

Opinion mining extracts people's own opinion using different methods such as questionnaire, form, emotions, views, etc. There are mixture of people with literacy and illiteracy staying in villages. To collect different views from different type of villagers, opinion mining uses different methods for them. After collecting views, these are converted into a particular structured format so that it will be very easy for analysis purpose. This job is performed by Hadoop tool which uses map reduce approach. This Hadoop uses distributed file system, i.e., HDFS which supports for storing both structured and unstructured data. Map Reduce is a parallel programming model for processing both structured and unstructured data through mapper and reducer class (Figs. 1 and 2).

4 Proposed Algorithm

In this section, we present the detailed Map Reduce implementation of result of opinion mining. After taking different opinions from villagers, then we use this Map Reduce approach on collected unstructured data set to get right conclusion. Following steps are:

1. To get response from villagers through feedback form and oral conversation.
2. Keep all opinions in digital form like notepad or MS Excel sheet.

Fig. 2 Opinion mining

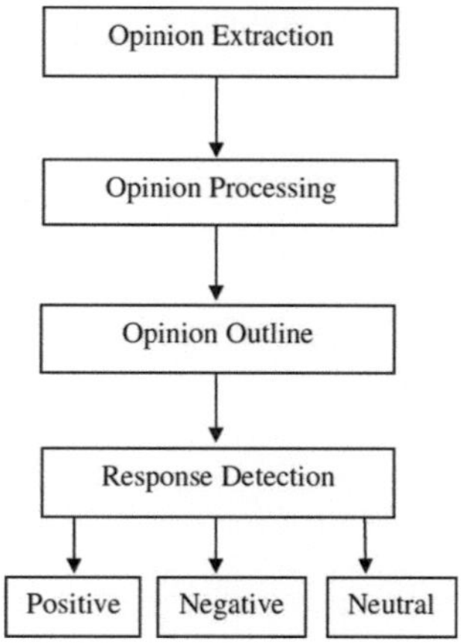

3. Applying opinion mining concept in digital form as per requirement to find smart village.
4. After step number 3, to make these required opinions in structured form, we apply map reduce approach.

Mapper Implementation

5. The input to the mapper is the unstructured data file. The record-reader reads the data file by breaking it into blocks of size nearly 50 MB.
6. After reading the blocks, the record-reader calls many mapper as many blocks read from the data file. The input to each invoked mapper is the block as value and the opinion data file name as key.
7. Each opinion unit or tokenized item is emitted from the mapper as key and one (intwritable) as its value.

In the mean phase, no combination is required. Every opinion unit data is emitted to the reducer to count its frequency. So the final output of the mapper is the opinion unit (key) and integer one (value) (Fig. 3).

Reducer Implementation

8. The reducer receives the key as opinion set and value as numerical one. So the number of reducers called is equal to the total set of opinion units in the opinion data file.
9. Then the final frequency table is prepared by summing the values for one key. The reducer receives the value as iterable so that it can be summed up.
10. After the summation of values for a particular key, it produces output.

The output is written to a file in HDFS (Fig. 4).

11. Then we filter this output of step number 10 using different decisions of villagers to satisfy the level of smart village.

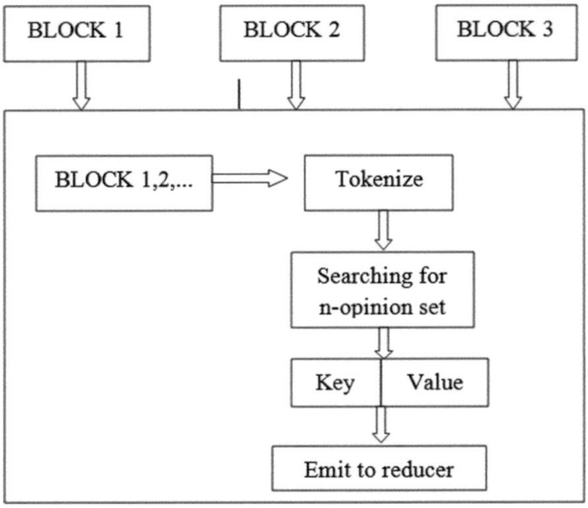

Fig. 3 Mapper phase

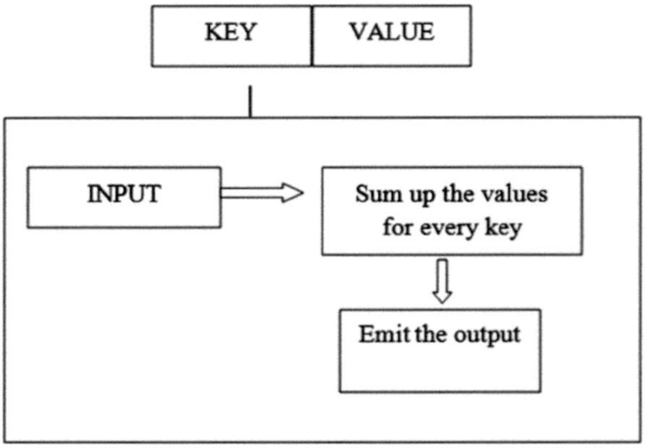

Fig. 4 Reducer phase

12. On the basis of satisfying criteria, we can conclude whether village can be fit for smart village or not.

5 Experimental Analysis

After getting output from map reduce approach, we get different opinions from villagers on different criteria.

Figure 5a tells about number of opinions by 100 no of villagers on 13 no of criteria for smart village. Figure 5b shows a pie-chart describing how many no. of

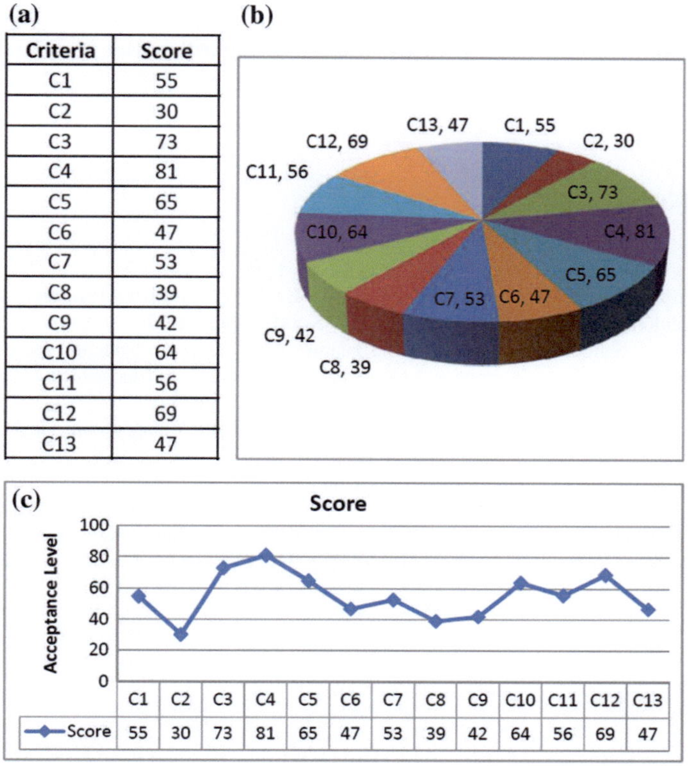

Criteria	Score
C1	55
C2	30
C3	73
C4	81
C5	65
C6	47
C7	53
C8	39
C9	42
C10	64
C11	56
C12	69
C13	47

Fig. 5 shows acceptance level of different criteria for smart village by villagers

villagers are agreed on 13 no of criteria. Figure 5c shows a graph between acceptance level of villagers and different criteria of the smart village. We can suggest to recommend smart village that among 13 criteria, at least 10 criteria should be accepted by villagers. Otherwise, we cannot recommend that village for becoming a smart village.

6 Conclusion

Opinion mining extracts variety opinions from villagers by above methods. To know how much response from villagers in different criteria for smart village, we use Hadoop tool which uses map reduce approach for calculating no of person's opinion. Here process of searching and finding no of opinions become fast by using map reduce approach so that we can take consideration into account quickly for recommending a particular village for smart village. In future, we can optimize our proposed map reduce method after applying opinion mining by GPU architecture.

Instead of using mapper function, we can use GPU architecture for mapping our collected opinions from villagers, by which our algorithm can take less time for execution than proposed algorithm.

References

1. http://www.smart.ap.gov.in/myvillage/manuals/smart.pdf
2. Kumar, P., Zaidi, A., Shrivastava, A., Jain, G.: Evolving sentiments towards e-governance using opinion mining. Int J Comput Appl (0975–8887), pp. 24–27
3. Sahoo, A.K., Sahoo, K.S., Tiwary, M.: Signature based malware detection for unstructured data in hadoop. In: 2014 International Conference on Advances in Electronics, Computers and Communications, INSPEC Accession Number: 14841793, pp. 1–6
4. http://www.easac.eu/fileadmin/PDF_s/reports_statements/Smart_Villages_concept_note_230113.pdf
5. http://ieee-smart-village.org/communities/asia/india/
6. https://cgspace.cgiar.org/bitstream/handle/10568/33322/CCAFSClimateSmartVillages2013.pdf
7. Nayak, A.S., Fernandes, R.: An study and analysis of opinion mining and sentiment analysis. Int J Adv Res Comput Eng Technol (IJARCET). **3**(12), pp. 4312–4315 Dec (2014)

Analyzing Online Groups or the Communities in Social Media Networks by Algorithmic Approach

Mamta Madan, Meenu Dave and Meenu Chopra

Abstract This paper focuses on communities or clusters which are the sets of nodes with lots of links within and very less to the outside of the network. The paper explains the concept of online generation communities and their framework in online social media networks (OSMNs), especially largest networking site, i.e., Facebook. There are many popular methods available for community identification like *Walktrap*, *Nibble*, *Label Propagation Algorithm* (*LPA*), *Fast Community Network Algorithm* (*FCNA*) which had been explored in the last decade. The *community framework* (*CF*) is the important and integral part of the OSMNs, but still we have to find the absolutely correct definition of the community in the real-world networks. In this paper, we try to give a correct definition of the community with its few important traits and from that we are able to recommend a different, simple and innovative framework (either flowchart or algorithm) which will resemble the real-world network. In our approach, we try to incorporate or consider those nodes which are overlapped in the community framework with the concept of shortest paths. We believe that our approach will be more favorable than other network methods which mostly generate the partitions.

Keywords Social media networks (SMNs) · Social network analysis (SNA) · Community identification (CI) · Graph clustering (GC) · Community framework (CF) · Modularity optimization (Q)

M. Madan (✉) · M. Chopra
GGSIP University, Delhi, India
e-mail: mamta.vips@gmail.com

M. Chopra
e-mail: meenu.mehta.20@gmail.com

M. Dave
JaganNath University, Jaipur, India
e-mail: meenu.s.dave@gmail.com

A.K. Saini et al. (eds.), *ICT Based Innovations*, Advances in Intelligent Systems and Computing 653, https://doi.org/10.1007/978-981-10-6602-3_12

1 Introduction

Before we define community, we should know about *network science* and types of networks, which includes *Sociology* (*social network analysis-SNA*), *Mathematics*, *Computer Science* (*Graphs*), *Economics*, *Bioinformatics* (*Networks*), and *Statistical Science* (*Complex Networks*). There are two types of networks (i) Dynamic network- that evolves over time for, e.g., Communication or social network (efficient and less time-consuming) and (ii) Static Network—evaluates the community framework from scratch for each change in network structure (long running time, same reaction to a small change).

What makes a community? It is a cohesive subgroup which has (i) *Mutuality of Ties*—which define everyone in the group has ties (edges) to one another, (ii) *Compactness*—Closeness reachability of the group with a small number of steps, (iii) *Density of edges*—high frequency of ties within the group, and (iv) *Separation*—*h*igher frequency of ties among group members compared to non-members.

It is an important challenge for the researchers to discover the community architecture from the different large, real and complex online networks. In [1–4], we discussed in detail about exploiting and mining social media platforms (SMNs) using predictive analysis through text, link, and spatial-temporal information especially Facebook networks. Many researchers during the last decade had developed several different algorithms to identify a community framework or structure in a complex network but have some drawbacks [5]. For example, few algorithms performed very badly on huge complex networks, some required few pre-requisites like community members, some were not able to detect the coinciding communities, some required few parameters to start the algorithm, few were unable to explore the mall or sparse classes (communities), some were specific to the domain, very less were able to work with only specific structures and only handful were able to generate good and stable partitions.

We have worked on an approach which is innovative that automatically identifies the total number of communities and will be able to find non-connected or secluded nodes even communities and finally divides the network into stable segments or partitions. We believe that with this approach or method we are able to produce different communities like coinciding, cohesive (less connected), small, and even sparse (scattered) communities. It is important to find or locate the smallest and sparse communities because their number is huge in real-work networks especially biological networks.

2 Literature Review

Today, most of the researchers related to network analysis are focusing on community identification, its structure and to its properties to give an exact picture of the community in real-world networks. One of the most popular methods is based on the *modularity optimization* or the [*Q*] factor [6–9] of a subgraph or subnetwork. Few approaches [6, 10–15] tried to compel each vertex to be allocated to one single community. This is not possible in real-world networks. Many algorithms have been developed, explored, and tested to detect the overlapping or coinciding communities [16–22, and 23]. Recently, to find link communities have been developed [20, 24, and 25]. Link communities help us to find coinciding communities as edges have their own unique identity as compared to nodes or vertices. Other approaches like statistical [26], static-informational approach [5, 27, 28], and dynamical or synchronization approaches [29–32].

Community identification (CI) is essentially clustering. But why researchers use CI in physical aspects rather than clustering because the "network big data" pose challenges to the old classical clustering method. Community identification can be defined as gathering of network vertices into groups in such a way nodes in each group is densely connected inside and sparser outside. The various applications are social network problems, Routing in Mobile Ad hoc networks, worm containments in cellular networks, etc. The process of discovering the community framework (CF) of a social media network is known as *community identification* (*CI*) or *it can define as gathering of network vertices into groups in such a way that nodes in each group are densely connected inside and sparser outside* (for example applications like social network problems, routing in mobile Ad hoc networks or Worm Containments in cellular Networks).

As we all know that real-time networks, i.e., online social media networks (OSMNs) are neither regular nor random in nature but also nontrivial in their topology, scale-free in nature and having universal properties like complex systems and present everywhere. The empirical network features which define real-world networks were discussed in [33] are; (a) *Power-law* (*heavy-tailed*) *degree distribution*, (b) *Small average distance (graph diameter*), (c) *Large clustering coefficient* (*transitivity*) (d) *Giant connected component*, *hierarchical structure*, etc. We also discussed various network analysis models in [33]; for example, (a) *Random graph model* (*Erdos & Renyi*), (b) *Preferential attachment model* (*Barabasi & Albert*), *and* (c) *Small world model* (*Watts & Strogatz*).

Network communities are groups of vertices which are similar to each other and *Community identification* (CI) is an assignment of vertices to communities. The non-coinciding communities are those where every vertex belongs to a single group. The aim of CI as well as *graph clustering* (*GC*) is same, i.e., partitioning network nodes into groups. In graph clustering (GC), the number of clusters is predefined or given as part of the input, whereas in CI the number of communities is typically unknown. Further community identification can be done in two ways (a) *Vertex clustering* (*vertex similarity*) *and* (b) *Graph partitioning* (*sparse cuts*).

In [94], it was explored that for any random graph, the degree of all the vertices (i.e., Distributions of edges) are either equal or homogeneous in nature because of the Poissonian distribution. But the real-world networks are not similar to random graphs, as they display big inhomogeneities. Also the degree of distribution is broad and scale-free it means many nodes with a lower degree and less number of nodes with higher degree exist in the network.

3 Algorithm Approaches

Before we discuss the algorithmic part of community identification, we want to throw some light on three basic and important properties [11, 13] which helped us in developing a framework to find communities. Those properties are (i) *community architecture*, (ii) *community (association) membership, and* (iii) *coinciding (overlapping) member properties, respectively*. The *first* property is used to describe the community; it is sub-network that connected densely internally within the members rather than externally. The strong ties community is one who has more connections internally than externally whereas weak ties communities are those which has more densely connected externally rather than internally. The *second* property is the community association or the membership property of a member which defines that the respective non-coinciding member has more neighbors within its own community rather than outside in any other community (i.e., Each node or vertex which is not coinciding with other communities should join the community which contains the maximum number of member neighbors). The *third* property is coinciding (overlapping) member properties that are used to represent the members of a community.

Further, there are two categories of coinciding vertices *first category*, depends upon the total number of connections between that vertex and corresponding communities and the shortest path exist between that vertex and the hub members (it consists of maximum number of neighbors within a community) of the corresponding communities, while the *second category* depends upon the community architecture or layout or topology. For the *first category* of coinciding nodes not only the number of connected links between the coinciding node and the matched communities should be identical, but also the path (should be equal and shortest) between that vertex and the core members of the comparable or matched communities. In the *second category*, the coinciding node should be strongly affiliated with the matched communities and also least number of connections between the two matched (corresponding) communities if such types of nodes have to remove.

We have explained the following methods or algorithms of community identification with their basic notion, constraints and their time complexity as mentioned in (Fig. 1).

I. *Edge—Betweenness*

Basic Notion: It is the improvised version of LPA [34]. This algorithm divides the network into subsequently smaller pieces by finding edges that “bridge” communities [35].

Constraints: Can be adapted to direct networks (digraph), Can be adapted to weights (no public software), do not able to draw conclusions about small communities (i.e., It gives a big picture).

Time Complexity: O (|V|^3) in general, O (|V|^2 log |V|) for special cases.

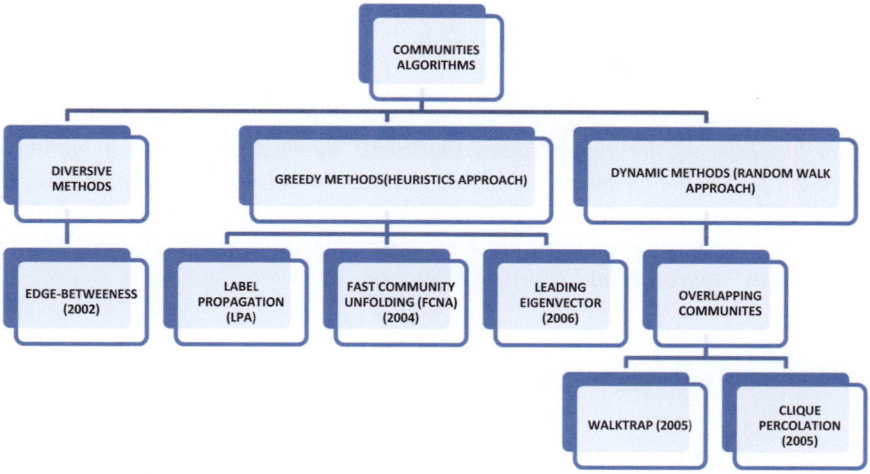

Fig. 1 Diagram to depict the taxonomy of community algorithms

II. *Fast Greedy*

Basic Notion: [34, 36–38] Try to randomly assemble larger and larger communities from the ground up. Start by placing each vertex in its own community and then combine communities that produce the best modularity at that step.

Constraints: Can be adapted to directed edges (digraph), Can be adapted to weights (no public software), it tends to aggressively create larger communities to the detriment of smaller communities.

Time Complexity: O (| E| V| log |V|) worst case.

III. *Leading Eigenvector*

Basic Notion: [39, 40] Use the sign on the components of the leading eigenvector of the Laplacian to sequentially divide the network.

Constraints: Can be adapted to directed edges (no public), Can be adapted to weights (digraph).

Time Complexity: O (| V| ^ 2).

IV. *Walktrap*

Basic Notion: [41] Simulate many short random walks on the network and compute pairwise similarity measures based on these walks. Use these similarity values to aggregate vertices into communities.

Constraints: Can be adapted to direct edges (digraph), can be adapted to weights, and can alter the resolution by walk length.

Time Complexity: depends on walk length O (|V|^2 log |V|) typically.

So far, none of the method discussed above shows the standard quantitative definition to find the community structure that can be acceptable in the real world. We should be able to develop the algorithm which is simple, stable enough to understand the partitions. It should have linear time complexity. It should be an unsupervised approach and heuristic in nature.

4 Improvised Algorithm

We propose that every algorithm must have following properties to detect community for a given graph (V, E) in online media networks.

1. **Scalable**, where time and space complexity are strictly sub-quadratic w.r.t. the number of nodes.
2. **Nonparametric**, where the number of communities need not be specified a priori.
3. **Consistent**, where effectiveness is consistently high across a page wide range of domains.
4. **Effective**, where global connectivity patterns are successfully factored into communities that are highly predictive of individual links and robust to small perturbations in the network structure.

The community identification is a challenging problem and for huge networks coming up day by day, it requires fast and accurate algorithms. The existing algorithm has a high order of time complexity. LPA [34] uses a *heuristic approach, linear time complexity, fast convergence, and can work on very large SMNs.* The algorithm works as follows:

Step 1 Let G = (V, E) be the given graph of N nodes.
Step 2 Initialize all nodes with a unique label Li.
Step 3 In each iteration, a node searches for the label maximally occurring in its neighborhood.
Step 4 It now changes to that label.
Step 5 Ties are broken uniformly randomly.
Step 6 Converge when all nodes have the same label as the one carried by a maximum of their neighbors

The algorithm discussed below has the same spirit as LPA (Label Propagation Algorithm). Our notion is based on the concept of community architecture which can be identified from subgraphs by evaluating the total count of in and out degree of each community, and it consists of following steps: (i) to load the adjacency matrix or lists with connected nodes or vertices, (ii) to locate–identify–remove strong and weak communities from adjacency lists, (iii) repeatedly assign and unassign nodes to detected communities depends upon the community membership property of the respective nodes, and (iv) continue the process until we get final identified communities and few isolated nodes.

Move 1: *Load the Adjacency List* with connected nodes or vertices.
if the list is empty, then got to RESULT/OUTPUT option (we get the isolated node), else Move 2:

Move 2: Locate, identify, and remove from adjacent lists *Weak Ties Communities*.

Move 3: Locate, Identify, and Remove from adjacent lists, *Strong Ties Communities*.

Move 4: *Select and Designate nonassigned* vertices to that Community in which its neighbors are present.

Move 5: *Discard the repetitious communities* or groups which are equal in number or proper subset of another Communities, in case of overlapping communities exist to remove the intersecting part of the smaller community.

Move 6: Accommodate all the missed out vertices or nodes to their specific community in which they are membership property belongs to.

Move 7: Authenticate various overlapping vertices, i.e., cross-check each and every overlapping vertex that exist between two communities. If vertex shows the community overlapping membership definitions, then put in the same, else allocate the vertex to the community which has the greater number of neighbors.

Move 8: Check identified communities (at least, if the community does not represent the definition of weak tie Community, then integrate it with the community which has the most populated connections).

Move 9: Look for any change required in each community? If it is true, then go to MOVE 6: Else next step.

Move 10: Look for any unassigned vertex or node? If it is False, then go to final result or output, else next step.

Move 11: Discard all the nonconnected components, i.e., Remove all the communities which are alone or isolated depends on the adjacency list of each unassigned node or vertex.

Move 12: We get all the identified communities as well as the isolated communities, if they are present.

4.1 Similarity Between LPA and Proposed Algorithm

The *greatest similarity* between both the algorithms is that both follow the same procedure, i.e., they both allocate each vertex in the network graph to a preferable community (generally which contains most numbers of its neighbors). Moreover, partitions generated by LPA and also by our algorithm, favor the same configuration.

4.2 Technical Difference Between LPA and Proposed Algorithm

The *technical difference* between our algorithm and LPA is that LPA loads the node communities, whereas our algorithm load connected edge or arc communities in the adjacency matrix or list (where each vertex corresponds to all its related edges or arcs). The LPA has some constraints like it requires some initial conditions, for its execution, it requires some tiebreak rules. It requires some randomly defined roots or seeds. Our approach tries to achieve near linear times complexity (i.e., *Deterministic approach*) for community identification which LPA is not able to achieve. The LPA looks for communities based on the process of propagation through the network by dynamically allocating the labels, while our algorithm works on static network layout or topology.

4.3 Possible Experiments Datasets—Facebook Higher Education (HE) Group

We still have to work on our algorithm execution phase, it is a just *a proposed or theoretical* model to detect not only communities but also the overlapping nodes. In the future, we will work, to evaluate the performance of our algorithm, with the LPA algorithm partitions. By literature review, we came to know that empirical networks can be evaluated by NMI (*normalized mutual information*) values, as discussed below, whereas many others can be evaluated by calculating the modularity Q metric, which is also mentioned below. As LPA works well on synthetic networks, so we expect our algorithm to also behave in the same manner, as both are based on the same notion. However, for empirical real-world networks, we believe that our algorithm will perform better than the other methods. Moreover, we expect higher stability and modularity. We conclude that our algorithm performs well and is competitive with other methods.

To detect communities, the possible datasets which we fetched in our previous work by using Network Analysis Software Applications (NASA) [42], from the largest social networking site, i.e., Facebook is having Higher Education (HE) Online Groups. It is a social network which is directed user–user friendships, network having 367 nodes, and 6468 vertices in Fig. 2 as compiled by us by using

NASA in January 2015. The Fig. 2 describes the graph network of the fetched online group, dataset and communities detected by NASA. In the future, we will try to compare and evaluate the strongly connected components (367) in Figs. (3 and 5) and weakly connected component (2) in Figs. (4 and 5) and total communities found (6) (Fig. 4) are detected by NASA with our proposed algorithm.

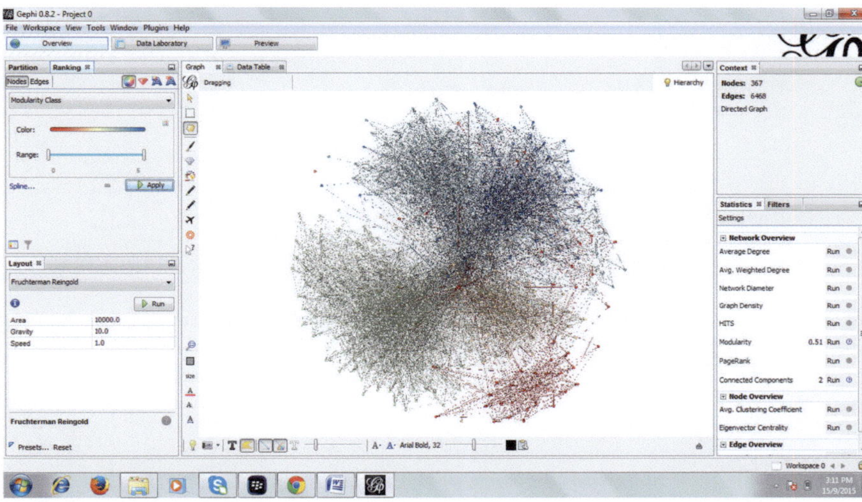

Fig. 2 The graph network for higher education group on Facebook with nodes 367 and edges 6468

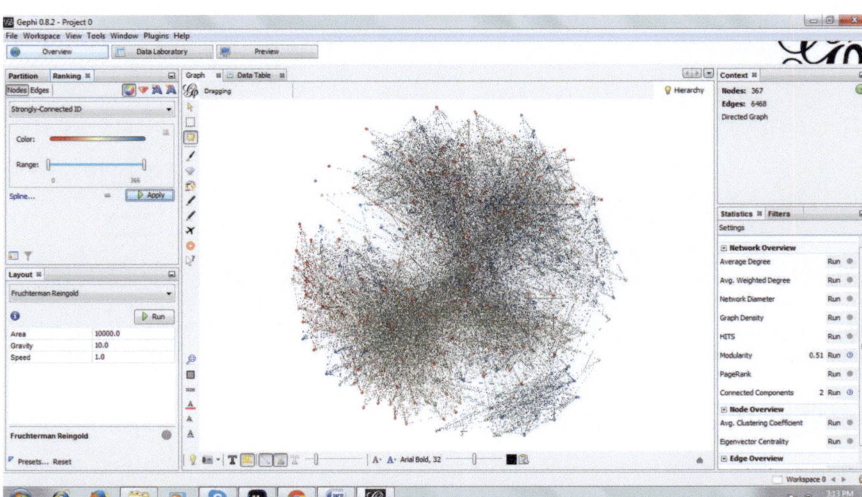

Fig. 3 Total number of strongly connected components ID which were found to be 367 for higher education group on Facebook

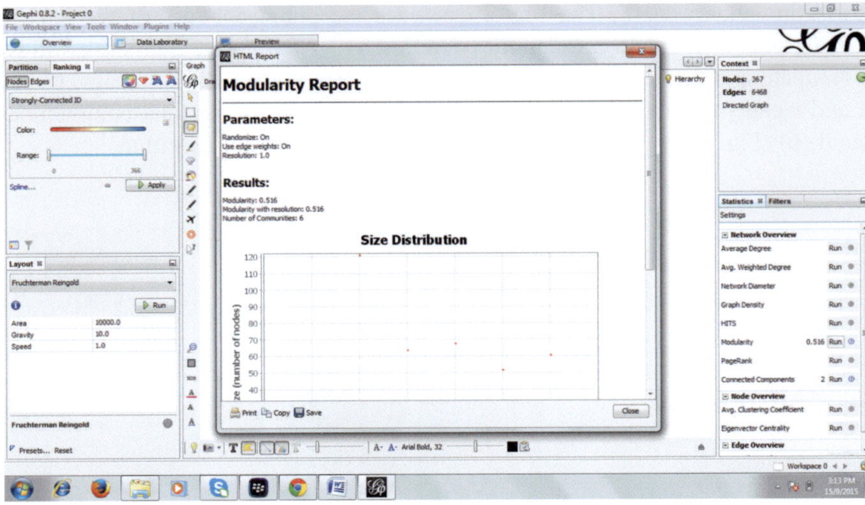

Fig. 4 Total number of communities as per modularity report, which were found to be 6 for higher education group on Facebook

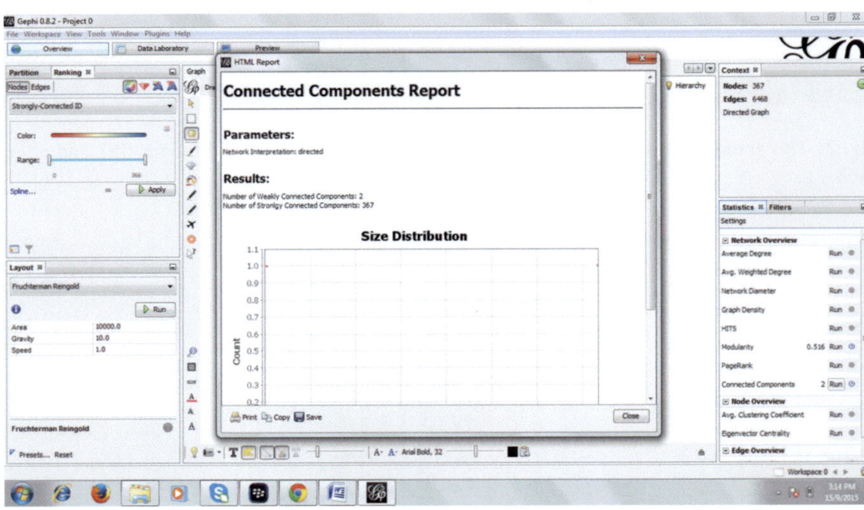

Fig. 5 Total number of strongly connected components which were found to be 367 and weakly connected components is 2 for higher education group

5 Algorithm Evaluation

At present, there is no universal technique to check the correctness of the segments created by the algorithms because it is a cumbersome process to predict the architecture of real-world networks beforehand. Now to evaluate quality of community segments, we are using the *(NMI) normalized mutual information measure* [16] which is defined below:

$$NMI(X|Y) = \frac{(H(X) + H(Y) - H(X, Y))}{(H(X) + H(Y))/2} \quad (1)$$

In above formula, *X* and *Y* represent reality and predicted communities, *H* (*X*) is the random entropy of *X*, and *H* (*X*, *Y*) *H* (*X*, *Y*) represents the joint entropy of *X* and *Y*.

$$Q = \sum (e_{ii} - a_i^2) \quad (2)$$

In other networks, we are using the above mathematical expression which calculates *modularity* [43] property to calculate the quality of a community segment (partition). It is generally believed that networks which are random in nature do not exhibit community architecture. To calculate modularity we have to define a matrix *e* where the each element *e {i, j}* depicts the fraction of the total number of connections between two contrasting communities, and the absolute fraction of ties particularly within the reach of a community it is *e {i, i}*. Finally, the sum of any row of e $a_i = \sum_j e_{ij}$ corresponds to the fraction of links connected to the community *i*, and the expected count of intra-community ties are just a_i^2. We can compare *e {i, i}* and a_i^2 directly, and a total of all the communities present in the real-world network.

Our motive is to develop an algorithm which has less time complexity, so that it can be used on large-scale complex real networks. It has two phases: (i) *to identify communities, and we expect its time-cost complexity which comes out to be O* (m^2 + *lmv*), *where v is the total count of vertices, m is the total count of identifying communities and l is the maximal size of the initial adjacency lists*, (ii) *after adjusting membership among communities, and we get O* (nm^2 + lm^2 + *mv*) *time-cost complexity, where n is the maximum number of coinciding nodes.* Therefore, the final calculated and expected time complexity of both phases that may come out to be $O((n + l)\ m^2 + lmv)$. In future, we are planning to use this algorithm on a huge real network due to its lesser time complexity.

6 Conclusion and Future Scope

Communities are not only groups of associated members which connect densely but also sparsely with the remaining part of the network. Many hidden patterns can be disclosed out by detecting the community and also by its architecture. But still, we have done a great deal of research to uncover the real-world systems whose structure is not being fully understood. With the help of various discussed approaches in the Sect. 3 of this paper, it is possible to throw some light on the layout or structure of these networks to improve, maintain, manage, control, and renovate them. In this paper, we propose a simple, different and innovative algorithm to identify a community framework of complex networks, especially social media networks based on network topology configuration. Comparing other popular approaches or algorithms discussed in the literature, we believe that our approach will work competitively on both artificial and real-world social media networks, but we know it is a long journey to get a breakthrough in community detection.

In the future, we still have to work on its practical execution, predictability and further its utility by identifying another category of coinciding vertices and explore the hierarchical architecture of the network. Presently our approach is to detect the most coinciding vertices where we are ignoring least coinciding (overlapping) vertices thereby making it difficult to divide larger communities into smaller ones. In the future, we may try to work on the optimization of modularity property of the network, which is an NP-hard problem.

References

1. Mamta, M., Meenu, C.: Social network wrappers (SNWs): An approach used for exploiting and mining social media platforms. Int. J. Comput. Appl. **97**(17), pp. 31–34 (2014). Source code available from http://research.ijcaonline.org/volume97/number17/pxc3897687.pdf
2. Mamta, M., Meenu, C.: To investigate relationships through text, link and spacial-temporal information in social media networks. Int. J. Sci. Technol. Res. (IJSTR), **4**(3) (2015). Source code available from http://www.ijstr.org/final-print/mar2015/A-Review-Paper-On-Exploring-Text-Link-And-Spacial-temporal-Information-In-Social-Media-Networks.pdf
3. Mamta, M., Meenu, C.: Using mining predict relationships on the social media network: Facebook (FB). Int. J. Adv. Res. Artif. Intell. (IJARAI), **4**(4) (2015). Source code available from http://dx.doi.org/10.14569/IJARAI.2015.040409
4. Mamta, M., Meenu, C.: Social media networks (SMN) an eye: to envision and extract information. Int. J. Adv. Res. Comput. Sci. Softw. Eng. **4**(2), (2014) Source code available from http://www.ijarcsse.com/docs/papers/Volume_4/2_February2014/V4I2-0462.pdf
5. Lancichinetti, A., Fortunato, S.: Community detection algorithms: a comparative analysis. Phys. Rev. E. 80, 056117 (2009)
6. Newman, M.E.J., Girvan, M.: Finding and evaluating community structure in networks. Phys. Rev. E. **69**, 026113 (2004)
7. Newman, M.E.J.: Modularity and community structure in networks. Proc. Natl. Acad. Sci. USA **103**, 8577–8582 (2006)
8. Duch, J., Arenas, A.: Community detection in complex networks using extremal optimization. Phys. Rev. E. **72**, 027104 (2005)
9. Blondel, V.D., Guillaume, J.L., Lambiotte, R. and Lefebvre, E.: Fast unfolding of communities in large networks. J. Stat. Mech. P10008 (2008)
10. Ravasz, E., Somera, A.L., Mongru, D.A., Oltvai, Z.N., Barabási, A.L.: Hierarchical organization of modularity in metabolic networks. Science **297**, 1551–1555 (2002)
11. Girvan, M., Newman, M.E.J.: Community structure in social and biological networks. Proc. Natl. Acad. Sci. USA **99**, 7821–7826 (2002)
12. McAuley, J., Leskovec, J.: Learning to discover social circles in ego networks. Adv. Neural. Inf. Process. Syst. **25**, 548–556 (2012)
13. Newman, M.E.J.: Communities, modules and large-scale structure in networks. Nat. Phys. **8**, 25–31 (2011)

14. Zhang, S., Zhao, H.: Community identification in networks with unbalanced structure. Phys. Rev. E. **85**, 066114 (2012)
15. Aldecoa, R., Marín, I.: Deciphering Network Community Structure by Surprise. PLoS One **6** (9), e24195 (2011)
16. Lancichinetti, A., Fortunato, S., Kertesz, J.: Detecting the overlapping and hierarchical community structure in complex networks. New J. Phys. 11, 033015 (2009)
17. Derenyi, I., Farkas, I., Palla, G., Vicsek, T.: Uncovering the overlapping community structure of complex networks in nature and society. Nature **435**, 814–818 (2005)
18. Gregory, S.: Finding overlapping communities in networks by label propagation. New J. Phys. **12**, 103018 (2010)
19. Lancichinetti, A., Fortunato, S.: Benchmarks for testing community detection algorithms on directed and weighted graphs with overlapping communities. Phys. Rev. E. 80, 06118 (2009)
20. Evans, T.S., Lambiotte, R.: Line graphs, link partitions, and overlapping communities.Phys. Rev. E. 80, 016105 (2009)
21. Gregory, S.: An algorithm to find overlapping community structure in networks. Lect. Notes Comput. Sci. **4702**, 91–102 (2007)
22. Shen, H., Cheng, X., Cai, K., Hu, M.B.: Detect overlapping and hierarchical community structure in networks. Phys. Stat, Mech. Appl. 388, 1706–1712, (2009)
23. Li, D., et al.: Synchronization interfaces and overlapping communities in complex networks. Phys. Rev. Lett. **101**, 168701 (2008)
24. Ahn, Y.Y., Bagrow, J.P., Lehmann, S.: Link communities reveal multi-scale complexity in networks. Nature 466, 761–764, (2010)
25. Evans, T.S., Lambiotte, R.: Link partitions and overlapping communities. Phys. Rev. E. **80**, 016105 (2009)
26. Ramasco, J., Fortunato, S., Lancichinetti, A., Radicchi, F.: Finding statistically significant communities in networks. PLoS One **6**, e1896 (2011)
27. Rosvall, M., Bergstrom, C.T.: Maps of random walks on complex networks reveal community structure. Proc. Natl. Acad. Sci. USA. 105, 1118–1123 (2008)
28. Rosvall, M., Bergstrom, C.T.: An information-theoretic framework for resolving community structure in complex networks. Proc. Natl. Acad. Sci. USA **104**, 7327–7331 (2007)
29. Boccaletti, S., Ivanchenko, M., Pluchino, A., Latora, V., Rapisarda, A.: Detecting complex network modularity by dynamical clustering. Phys. Rev. E. **75**, 045102(R) (2007)
30. Oh, E., Choi, C., Kahng, B., Kim, D.: Modular synchronization in complex networks with a gauge Kuramoto model. Europhys. Lett. 83, 68003 (2008)
31. Lancichinetti, A., Fortunato, S.: Consensus clustering in complex networks. Sci. Rep. 2, srep 00336 (2012)
32. De Los Rios, P., Gfeller, D.: Synchronization and spectral coarse graining in oscillator networks. Phys. Rev. Lett. **100**, 174104 (2008)
33. Mamta, M., Meenu, C.:Network analysis by using various models of the online social media networks. Int J Adv Res Comput Sci. **6**(1), pp. 111–116 (2015)
34. Raghavan, U., Albert, R., Kumara, S.: Near linear time algorithm to detect community structures in large-scale networks. Phys. Rev. E **76**(3), 036106, 87–88 (2007)
35. Girvan, Newman.: Community structure in social and biological networks. PNAS. (2002)
36. Fast algorithm for detecting community structure in networks. Phys. Rev. E. (2004)
37. Clauset, Newman, Moore.: Finding community structure in very large networks. Phys. Rev. E. (2004)
38. Wakita, Tsurumi.: Finding community structure in mega-scale social networks. (2007)
39. Newman.: Finding community structure in networks using the eigenvectors of matrices. Phys. Rev. E. (2006)
40. Leicht, Newman.: Community structure in directed networks. Phys. Rev. Le. (2008)
41. Pons, Latapy.: Computing communities in large networks using random walks. JGAA. (2006)
42. Mamta, M., Meenu, D., Meenu, C.: Social network analysis (SNA): In Facebook higher education groups through NASA (Network Analysis Software Applications), Int J Artif Intell Knowl Disc (IJAIKD), ISSN 2231–0312

43. Danon, L., Diaz-Guilera, A., Duch, J., Arenas, A.: Comparing community structure identification. J. Stat. Mech. Theor. Exp. P09008 (2005)
44. Zhou, C., Yuan, W.J.: Interplay between structure and dynamics in adaptive complex networks: Emergence and amplification of modularity by adaptive dynamics. Phys. Rev. E. **84**, 016116 (2011)
45. Mamta, M., Meenu, C.: Social media networks (SMN) an eye: to envision and extract information. Int. J. Adv. Res. Comput. Sci. Softw. Eng. **4**(2), (2014)

Open Source EJBCA Public Key Infrastructure for e-Governance Enabled Software Systems in RRCAT

Alok Jain, Sarthak Gupta, Mangalesh Vyas, Diptikant Pathy, Gitika Khare, Alpana Rajan and Anil Rawat

Abstract Deployment of paperless, workflow-driven software systems require high level of security and sophisticated authentication techniques. Any security solution that is integrated with digital applications need to answer the fundamental question of establishing *authenticity, access control, nonrepudiation* and *data integrity*. A major concern in transactions carried out in e-governance systems is the need for replacement of ink based or "wet" signature with an e-signature implemented using Digital Signature Certificate (DSC). Also, as any web-enabled application or service is prone to a multitude of attacks, any discussion on e-Governance enabled software system is incomplete without consideration of "security" as a prominent aspect of "e-signatures." An e-signature may be considered as a type of electronic authentication which can be applied to online transactions. There are a number of technologies, products and solutions for securing electronic infrastructure of an organization. It must be taken care that the level of sophistication provided by the chosen security solution should commensurate with level of security required by applications being designed. To operate critical web-enabled applications, a public certificate-based solution provided by a robust, enterprise grade PKI (Public Key Infrastructure) is one of the most widely deployed, standard solution. PKI-based security solutions can provide a high level of assurance to all application domains, e.g., digital form signing, business process automation, etc. PKI is a consistently evolving security process with a number of applications in government as well as business domains. It is one of the most appropriate security mechanisms for securing data, identifying users, and establishing a chain of trust to secure electronic documents. This paper discusses our experience in setting up and implementing the open source EJB Certifying Authority software, accompanying challenges, along with advantages and limita-

A. Jain (✉) · S. Gupta · M. Vyas · D. Pathy · G. Khare · A. Rajan · A. Rawat
Computer Division, Raja Ramanna Centre for Advanced Technology, Indore, India
e-mail: alok@rrcat.gov.in

A.K. Saini et al. (eds.), *ICT Based Innovations*, Advances in Intelligent Systems and Computing 653, https://doi.org/10.1007/978-981-10-6602-3_13

tions of implementing an in-house PKI setup and integrating this PKI functionality with other in-house developed paperless, workflow-driven software systems.

Keywords PKI · Digital signature certificate · Workflow · Digital signing · EJB certifying authority

1 Introduction

Traditional workflow in RRCAT is accomplished by obtaining ink or "wet" signature at each and every step due to statutory government requirements. However, traditional workflows are paper based which makes them prone to various losses and costs such as loss of documents, loss of productivity due to employee time spent in managing papers and files, increased administrative costs involved in sharing and distribution of paper copies, backup and storage costs, etc.

In computer-based workflow, information is easily and speedily transferred from one location to other as compared to information that is moved in form of a printed paper. Transitioning to a "paperless" workflow can accord significant benefits in terms of costs saved due to electronic filing of documents, low cost high density archiving capabilities, easier backup, easier searching, sharing and distribution capabilities. However, paperless workflow raises many security challenges:

- Authentication—how to establish user's identity.
- Access Control—how to ensure that only an authorized user is able to see a particular document or use a service.
- Nonrepudiation—how to establish the legality of a completed transaction and ensure its nonrevocation.
- Integrity—how to ensure the accuracy and consistency of a document over its entire lifetime.

To solve aforementioned challenges, a public key certificate-based system has been proposed. Public key certificate is based on Public key cryptography (also called asymmetric-key cryptography) which uses a pair of keys with one public and one private key. These keys are mathematically related to each other. Possession of private key and its usage in a transaction leads to *authentication* of an entity and by default establishes *nonrepudiation,* since by design that entity can be the sole possessor of its own private key. Corresponding public key is distributed to those parties that intend to participate in a transaction with the said entity.

However, any party involved in a transaction with an entity, shall have one major concern regarding the public key of the entity, i.e., it is not possible to know for sure that the key being used by the entity for a transaction actually belonged to that entity. There can be a possibility that another entity monitoring the said transaction between sender and receiver changed the original key to a substitute key. Establishing *access control* and data *integrity* require more elaborate arrangements.

To address aforementioned issues, the concept of PKI has been formulated. A typical PKI system contains a trusted third party that vouches for the identity of the owner of a public key. This trusted party, known as a Certification Authority (CA), typically accomplishes this by creating and issuing a digital certificate. This certificate acts as a link between public key and a particular entity. The certificate contains public key and information of the entity.

The CA signs this digital certificate with its own private key. Availability of CA's signature enables any third party to use the CA's public key to verify authenticity of the signature and therefore establish validity of the certificate. Once the certificate has been validated, the owner's public key (published in the digital certificate) can be used with confidence to carry out transactions with the entity owning the certificate.

2 PKI Setup at RRCAT

Enterprise Java Bean Certificate Authority, or EJBCA, has been used as the basis of PKI setup at RRCAT. EJBCA is an open source software and its source code is published under the Lesser GNU General Public License (LGPL) [2].

The EJBCA system is written in Java EE which makes it platform independent. It has been designed to be highly scalable. EJBCA allows interfacing with multiple databases, both open source as well as commercial. EJBCA maintains logs for every step being taken both at the user as well as administrator level. It allows publishing the certificates in LDAP directory [2].

EJBCA implements almost all of the standards developed in the field of PKI, e.g., X.509, OCSP and TSP, etc. Furthermore, EJBCA supports various asymmetric encryption as well as modern hash algorithms [2].

EJBCA PKI consists of software modules that implement the entire PKI functionality, i.e., the CA (along with inbuilt Registration Authority), services that take care of tasks like entity enrolment, entity certificate renewal, availability of programming interfaces for directory services and third party applications, private-key management, et al. EJBCA supports strong security and high availability [2].

EJBCA PKI supports open standards of certificates like X509 v3, PKCS, PEM, DER, etc., and thus it assures interoperability with other open technology platform to provide end-to-end PKI solution. EJBCA also provides easy to use web-based user and administrative interfaces. It provides robust infrastructure for carrying out complete lifecycle management of digital certificates starting from generation, distribution, renewal to their eventual expiry. Also, it allows us to enforce various policies related to certificate strength and validity period as specified by Government of India from time to time.

3 PKI Workflow and Lifecycle

The EJBCA end-user interface has been configured and administrative interface has been secured (Fig. 1).

The End-User Interface can perform the following tasks:

- Submit a certificate request
- Import a certificate into a digital wallet
- Submit a Server/PC certificate request

The Administrative User Interface can perform the following tasks:

- Approve, reject, revoke certificates
- Update and generate Certificate Revocation List (CRL)
- Manage policies, specifying key strengths, etc.

Following is the implemented Certificate Practise Statement [3]:

- RSA Key Constraints: For employees, 2048 bits key and for servers, 4096 bits key strength has been imposed according to current guidelines.
- Validity Rule: For employee—5 years and for server—10 years validity has been enforced.

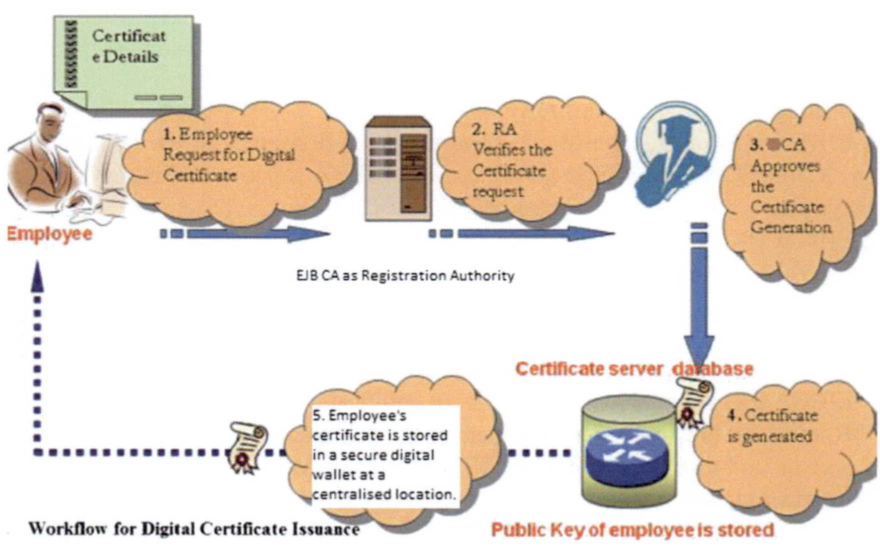

Fig. 1 Workflow for digital signature certificate issuance

- Unique Certificate Constraints: Single issue of certificate to the same employee/server is enforced.

Using the EJBCA PKI setup, Digital Certificates have been implemented for robust two factor authentication of the entity—proof of possession of private key by the entity and validation of public key belonging to the particular entity by EJBCA.

4 Implementation of Digital Certificate-Based Workflow-Driven Software Systems

Traditional workflow in RRCAT is accomplished by obtaining physical signature at each and every step. In computer-based workflow systems, information is easily and speedily transported from one location to another. However, data and documents, which are considered confidential and restricted (for use only by authorized users), exchanged using digital workflow pose a serious disadvantage since it is easy to modify the data/document. It is also possible that the digital workflow itself may be manipulated.

In addition to PKI system's ability to establish user identity, it also enables one to "digitally sign" electronic documents. The process of digitally signing a document consists of computing a hash of the contents of the document, encrypting the hash with the signer's private key and attaching the encrypted hash (signature) to the document. Since the sole possessor of a private key can perform this signing operation, this procedure mirrors the traditional workflow, i.e., an entity who puts a physical "wet" signature on a document is the only one who can put that written signature, as s/he is the sole possessor of that particular signature. Verification of the PKI digital signature can be carried out by decrypting the hash using the signer's public key and matching this hash against the recomputed hash of the received document's content portion. This is on the same lines as anyone can verify a written signature, establishing *authentication* in our software application.

This technology enables us to develop capability to carry out paperless transactions that need approvals from individuals, with them mirroring a paper-based traditional workflow. The signing authority carries out the electronic transaction, putting its "signature" electronically. Signature can be verified and action can be taken accordingly in further steps of the workflow. This enables *nonrepudiation* in our system, since only the sole owner of a private key can put his mark (signature) on a document. A hash of required data is computed and stored so as to allow for checking the *integrity* of required data.

Figure 2 shows snap shot of the administration interface for setting properties of the certificates.

Registered user's data has been provisioned in LDAP using email-id as unique DN, thereby user can generate and save certificate in browser store of his PC. This provides appropriate *access control* to the users. Figure 3 shows generation of DSC.

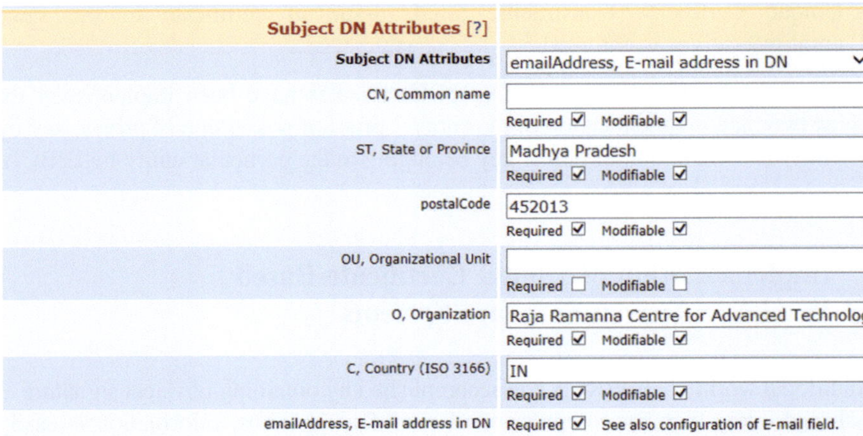

Fig. 2 Setting properties of the certificate

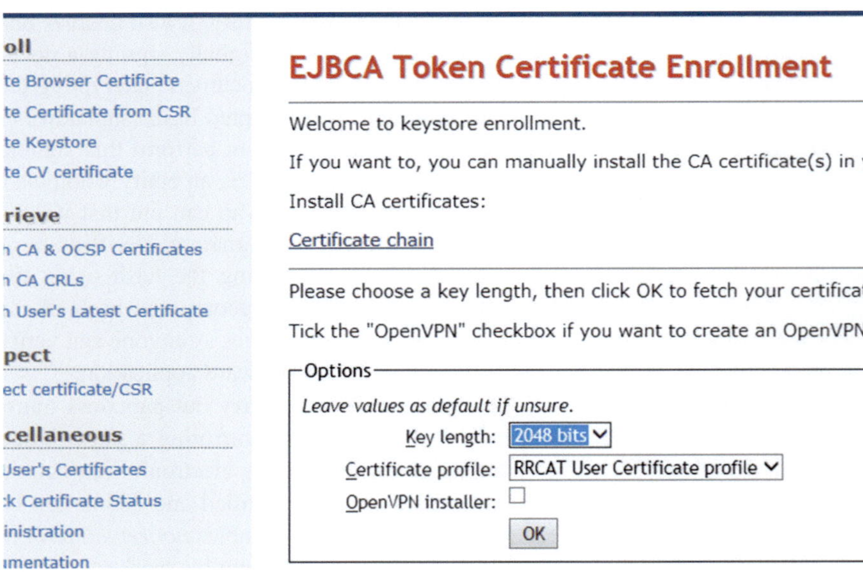

Fig. 3 Generation of digital signature certificates (DSC)

Work flow-based applications have been developed and implemented on RRCATInfonet—RRCAT Information Portal to move toward paperless transactions leading to implementation of e-governance applications in RRCAT.

Some of the workflow-based systems where Digital Signature Certificate (DSC) based on PKI technology has been enabled are Leave Request and Approval Workflow Software, Visitor Entry Requisition and Approval System and Material Gate Pass Preparation and Approval Workflow Software.

In all the PKI-enabled software systems, PDF file containing the details of the transaction is generated. Details of date and time stamp of the underlying transaction are stored. Email notification is send to concerned authorities for approval. Approval is carried out by authorities using DSC issued to them. After approval, appropriate information which can be used to vouch for credentials of aforementioned authorities are stamped in the PDF file at appropriate places. Critical information pertaining to the carried out transaction is logged in a database for audit purposes.

Figure 4 shows logging facility available in EJBCA. This facilitates validation of any digitally signed document even at a later stage.

Issued DSCs belonging to users are placed in a secured object on a centralised server, with users having passwords to access those objects. Centralized setup is a design decision so as to allow ease of use for the users, since they are not required

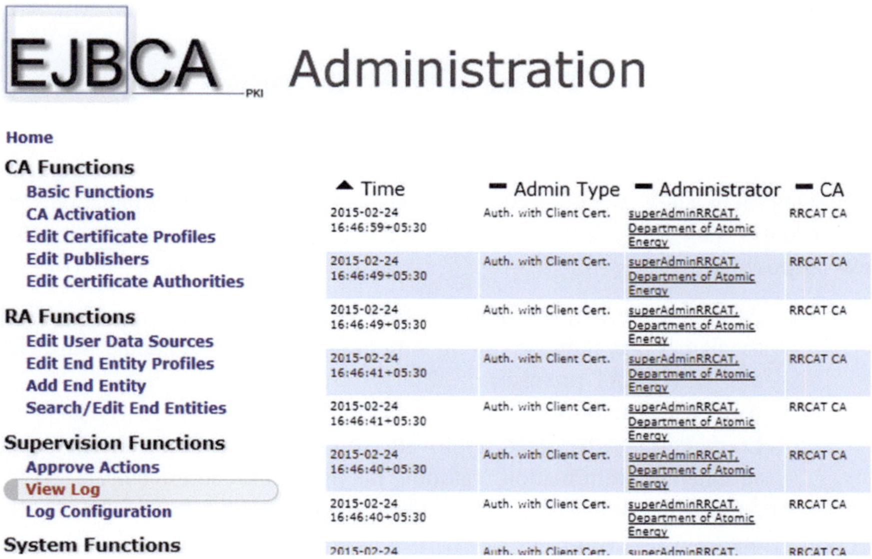

Fig. 4 Logging facilities in EJBCA

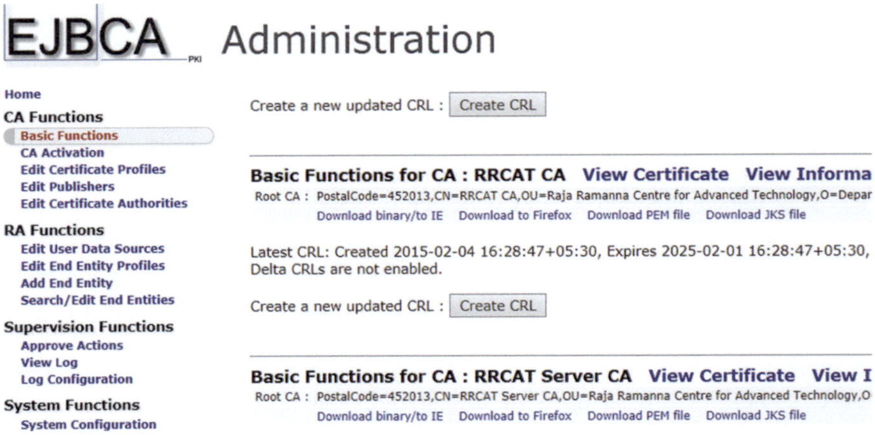

Fig. 5 Generation of CRL

to carry their DSCs in physical media with them for using it from various locations (office/meeting rooms, etc.). This has provided location transparency to the users as far as use of DSC is concerned.

CRL has been provisioned for validating expiry of user certificates. PKI-enabled applications can access it. Figure 5 shows generation of CRL.

5 Typical Flow of a Paperless Workflow-Driven Software System

Following steps show workflow of a typical paperless software application. We have taken Visitor Entry Permit Requisition and Approval System for demonstration purpose (Fig. 6):

Step 1: An employee prepares Entry Permit Requisition form, which contains the names and addresses along with purpose of visit of people who wish to visit RRCAT premises.

Step 2: Approving Authority logs in the system and views the newly prepared Entry Permits that need approval. Approving authority can also view summarized information regarding his last transaction details.

Step 3: Approving Authority enters his/her passphrase to select his/her secure digital wallet (Fig. 7).

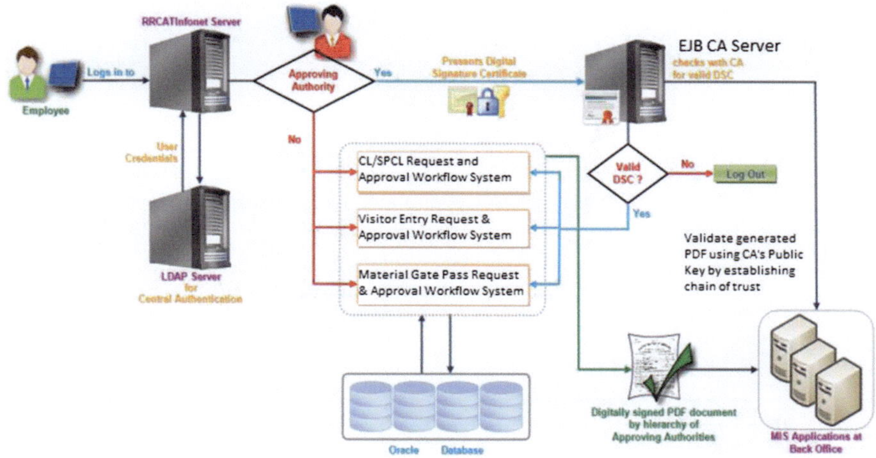

Fig. 6 Schematic view of workflow-based application using PKI at RRCAT

Fig. 7 Approving authority's passphrase entry screen

Step 4: If all of the supplied information is correct, Approving Authority's DSC is applied to the PDF Document of Entry Permit.

Step 5: A signed Entry Permit document is displayed to the employee who prepared it (Fig. 8).

Step 6: To verify the details of a signed document, verification is performed programmatically and shown to the concerned party (Fig. 9).

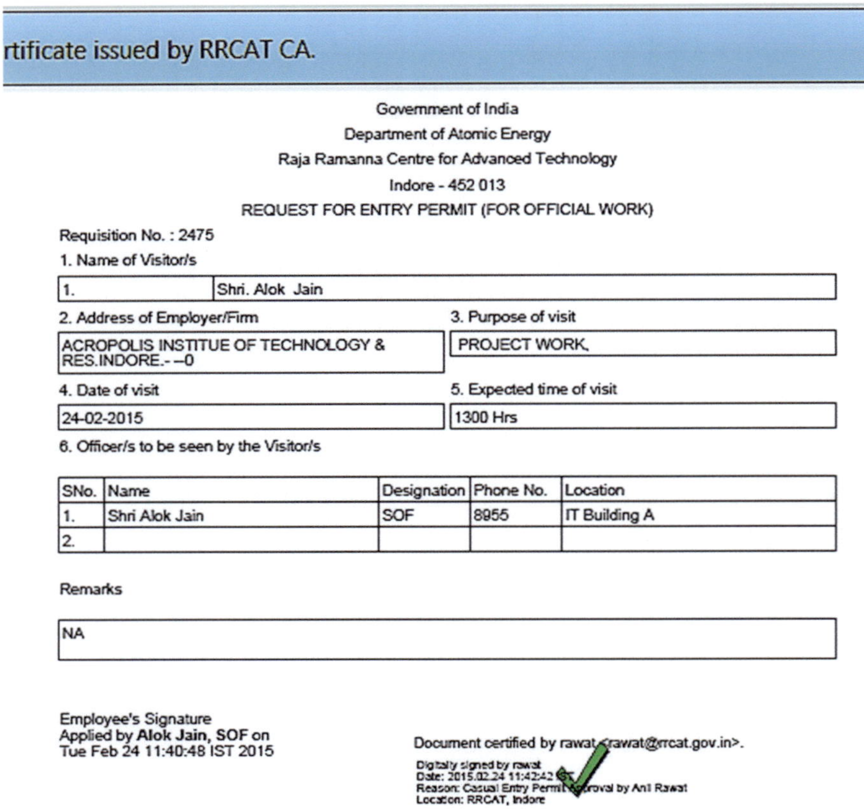

rtificate issued by RRCAT CA.

Government of India
Department of Atomic Energy
Raja Ramanna Centre for Advanced Technology
Indore - 452 013
REQUEST FOR ENTRY PERMIT (FOR OFFICIAL WORK)

Requisition No. : 2475

1. Name of Visitor/s

1.	Shri. Alok Jain

2. Address of Employer/Firm: ACROPOLIS INSTITUE OF TECHNOLOGY & RES.INDORE.- –0

3. Purpose of visit: PROJECT WORK,

4. Date of visit: 24-02-2015

5. Expected time of visit: 1300 Hrs

6. Officer/s to be seen by the Visitor/s

SNo.	Name	Designation	Phone No.	Location
1.	Shri Alok Jain	SOF	8955	IT Building A
2.				

Remarks

NA

Employee's Signature
Applied by **Alok Jain, SOF** on
Tue Feb 24 11:40:48 IST 2015

Document certified by rawat <rawat@rrcat.gov.in>.
Digitally signed by rawat
Date: 2015.02.24 11:42:42 IST
Reason: Casual Entry Permit Approval by Anil Rawat
Location: RRCAT, Indore

Fig. 8 A signed entry permit document

Verify Details for "1461_V1_800DF07C_20150225120356.pdf"	
Status	Complete
Field name	SignField1
Signer name	rawat
Signature covers whole document	true
Subject	{O=[Department of Atomic Energy] OU=[Raja Ramanna Centre for Advanced Technology] CN=[rawat] C=[IN] ST=[Madhya Pradesh] E=[rawat@rrcat.gov.in]}
Signature date	Tue Jan 27 13:48:00 IST 2015
Document modified	false

Fig. 9 Verification details of a signed document

6 Advantages of Setting Up In-House Open Source PKI

Setting up in-house PKI system accords following benefits:

- Total Cost of Ownership—Opting for an external PKI provider would have lead to significant monetary costs owing to certificate acquisition and their subsequent renewal, along with overhead time costs related to inherent delays in lifecycle management of certificates due to dependence over an external PKI provider.
- Domain Segregation—Software applications in RRCAT are on intranet domain which is completely segregated from Internet, i.e., applications hosted over Intranet cannot access Internet under any circumstance. Dependence over an external PKI provider would require opening up intranet to allow applications to communicate with external PKI's services.
- External Audit Requirements—In-house PKI system allows us to generate audit trails of signed documents as per the statuary requirements of Government of India.
- Open Source—EJBCA PKI is available under open source LGPL. The open source nature of software has allowed us to customize and install the software as per our requirements.
- Key Escrow—EJB CA supports key escrow feature which is a system in which the keys required for decrypting the previously encrypted and stored data are held in a backup source so that, if required, Government of India may get access to those keys and consequently the original material.
- Manageability—Using LDAP, EJB Certificate Authority helps us to meet Certificate provisioning needs of our organization. Web-based user and administrative interfaces provide elegant solution for organizational deployment.

In-house developed software system accord following advantages:

- Allows users to sign documents without requiring any extra software installed in their computers, except web browser.
- Users would not be tied to their local machines; they will be able to sign on-the-go, as long as they have access to the web application.
- The digital signature is directly embedded into the document. This allows the full document to serve as a self-contained e-record. Verification of the signature and integrity of contents can be carried out anytime.
- The hash algorithm generates a unique digital fingerprint of the document. Even a slight modification carried out in the document shall lead to a different digital fingerprint.

7 Limitations of Implementing In-House PKI Setup

There are certain limitations that we have observed while setting up EJBCA-based PKI setup at RRCAT:

- For high availability and scalability, EJBCA setup has to be replicated on different set of servers.
- CRL services are currently offered in off-line mode. Proper Online Certificate Status Protocol (OCSP) service setup needs to be done.
- Time-Stamping service is currently unavailable.
- Lack of dedicated support for any deployment and configuration issues if commercial services have not been subscribed to.

8 Conclusion

In-house design, development and commissioning of aforementioned setup has accorded a robust and standards compliant facility to RRCAT for digitizing and automating various workflows. Applications that require signature-based approvals are greatly benefited since requirement to print, manage and preserve paper-based documents, which leads to slower and error prone process, are virtually eliminated. Applications that have been made PKI enabled have led to improvements in user experience along with shortening the process times leading to considerable reductions in administrative costs while simultaneously contributing toward "Digital India" program of Government of India.

9 Future Work

Following tasks can be performed to enhance the supplied functionality:

- Time Stamping and OCSP service configuration.
- Replication of entire EJBCA setup to ensure high availability.

References

1. Forouzan, B.A., Mukhopadhyay, D.: Cryptography and Network Security, 2nd edn. New Delhi, India, Tata McGraw Hill Education Private Limited (2012)
2. EJBCA—Open Source PKI Certificate Authority, (Online). Available: http://www.ejbca.org/docs/index.html

3. Stallings, William: Cryptography and Network Security, 4th edn. Pearson Education (India), New Delhi (2009)
4. Jarupunphol, P., Mitchell, C.: PKI implementation issues in B2B ECommerce. In: Gattiker, U. E. (ed.), EICAR Conference Best Paper Proceedings (ISBN: 87-987271-2-5) 14 pages. EICAR, Copenhagen (2003)
5. Kiran, S., Lareau, P., Lloyd, S.: PKI- A Technical Perspective, (Online). Available: http:// www.oasis-pki.org/pdfs/PKI_Basics-A_technical_perspective.pdf
6. Java Platform, Standard Edition 6 API Specification, (Online). Available: http:// docs.oracle. com/javase/6/docs/api/
7. IText, Programmable PDF Software, (Online). Available: http://www.itextpdf.com Available: http://www.itextpdf.com/

Anticipation of Gross Domestic Product Using World Development Indicators

Kavita Pabreja

Abstract Gross Domestic Product (GDP) is one of the most important instruments to understand the economical position of an economy. The economic health of a country depends upon many factors, namely. consumption, business investment, government expenditure and net exports. The purpose of this study is to find out correlation among health, climate, and education related indicators of various countries as per their development status. The selected and reduced subset of indicators has been used for forecasting of GDP corresponding to High Income and Upper Middle Income countries using the World Bank's collection of indicators, assembled from officially documented international sources.

Keywords Gross domestic product · Health · Education · Development · Correlation · Artificial neural network

1 Introduction

The economic development of any nation is closely associated with the increase in the utilization and burning of fossil fuels, coal, oil, and natural gas by factories and electric power plants, motor vehicles, and family units. The consequential carbon dioxide (CO_2) emissions have turned into the largest source of greenhouse gases that do not allow the infrared radiation from the earth to leave the atmosphere and build the risk of global warming [1]. It has been suggested that every country must put in efforts to reduce the CO_2 emissions for the sake of its citizens.

These emissions affect the agriculture directly that in turn brings adverse effects on the economy of a country. As concluded by Smith [2], a little enhancement in average temperature worldwide (approximate change of 2 °C, calculated with reference to year 1990 temperature readings) would cause in net negative impacts

K. Pabreja (✉)
Maharaja Surajmal Institute (Affiliated to Guru Gobind Singh Indraprastha University), New Delhi, India
e-mail: kavita_pabreja@rediffmail.com

A.K. Saini et al. (eds.), *ICT Based Innovations*, Advances in Intelligent Systems and Computing 653, https://doi.org/10.1007/978-981-10-6602-3_14

on GDP of various developing countries and net positive market impacts on GDP of numerous developed economies. As a result, this would increase gaps in the income disparity among countries across the globe.

Health of people of a country plays a significant role in its growth and development. The importance of Adult Survival Rate on growth rates for poor countries has been discussed by [3]. The authors discussed that other parameters of health such as disease occurrence rates and cognitive functioning are imperative for sustaining a balanced provision of trained manpower which is a vital component for growth of economy.

In another study, the authors have examined the role of health to economic growth [4] and concluded that good health effects the economic growth of a nation in a constructive and numerically significant way.

Education is one of the most dominant mechanism for decreasing poverty and inequality and lays a basis for continual economic growth. The effect of education on GDP has been studied by [5]. The authors argued that the skills available in the labor force and the value of those skills are important determinants of economic performance of any country. There is requirement of workers with higher levels of education in order to handle complex services and production systems [6].

In the past, different Soft Computing techniques have been applied by many researchers for forecasting GDP of a nation. Neural networks have been applied for forecasting of macro-economical variables and an evaluation of different linear and nonlinear models has been done [7]. The authors have found that multivariate linear models are better. Different variants of neural networks have been used for anticipating the future of Egypt's cereal and it was concluded that ANN performed better than ARIMA [8]. GDP of Britain has been forecasted by authors using ANN and they have compared two different training algorithms [9]. The forecasting ability of ANN to anticipate financial output increase based on monetary and financial variables has been found to be better than linear models, as discussed in [10]. The GDP of Malaysia has been forecasted based on various economical indicators [11]. The authors have also compared econometric approaches with ANN and have demonstrated that ANN has better performance. A combination of ANN and ARIMA has also been experimented and in comparison with ANN or ARIMA, this hybrid model has produced more convincing results [12].

Most of these studies are based on either statistical relationship between different indicators or application of soft computing techniques using economical indicators as input variables. This paper differs from all of these studies as it is based not only on economical indicators but also on Education statistics and Health, Nutrition and population statistics. Also this piece of research is not just for a single nation but for different economical segments globally. In this paper, we have analyzed the most recent and precise global development datasets accessible at The World Bank's official website and using strong correlations as the basis of selection of a small subset of features, we have forecasted the GDP of countries in UMI and HI categories. Also, a few unusual correlations have been observed that are discussed in Results section. The paper is arranged as follows: Sect. 2 discusses Data and Methodology including data collection, countries under study, indicators used for the analysis and feature reduction. Section 3 explains the concept of Artificial Neural Network used for forecasting GDP, Results and Discussions are elaborated in Sect. 4 followed by conclusions as Sect. 5.

2 Data and Methodology

Datasets utilized for the study have been downloaded from The World Bank website. Various steps followed, namely selection of countries, different indicators responsible for economic growth and development, preprocessing of datasets and finally reduction of datasets, have been explained in following subsections.

2.1 Data Collection

There is huge data available for download for 249 countries, for 1343 attributes for 55 years for the time period 1960 till date [6]. The countries as per their development status have been categorized by The World Bank in four groups: High Income group, Upper Middle Income group, Lower Middle Income group and Low Income group.

2.2 Countries Under Study

Under High Income Group, the data corresponding to Singapore, France, USA, UK, Japan, Germany and Kuwait; in Upper Middle Income group, the data for China, Brazil, Iraq, Malaysia, Mexico. South Africa and Thailand has been collected for the study.

2.3 Indicators Under Study

In this paper, data related to three important determinants of GDP have been selected. The records corresponding to attributes describing the World Development Indicators; Education statistics; Health, Nutrition and population statistics have been collected for the time period 1980–2014 for the above stated two groups of economies [6]. The list of attributes under consideration, under each of these two groups has been explained below.

2.3.1 World Development Indicators

World Development Indicators (WDI) is the key World Bank repository of development data, assembled from officially distinguished global resources. It includes the latest and precise global development data, at national, regional and global level. The indicators considered under the study are as follows: Agriculture, value added (% of GDP); CO_2 emissions (metric tons per capita); Domestic credit provided by financial sector (% of GDP); Electric power expenditure (kWh per capita); Energy utilization (kg of oil equivalent per capita); Exports of goods and services (% of GDP); External liability stocks, total (DOD, current US$); Foreign direct investment, net inflows (BoP, current US$); GDP (current US$); GDP growth (annual %); GNI per capita, Atlas method (current US$); GNI, Atlas method (current US$); Gross capital formation (% of GDP); Imports of goods and services (% of GDP); Inflation, GDP deflator (annual %); Internet users (per 100 people); Life expectancy at birth, total (years); Merchandise trade (% of GDP); Mobile cellular subscriptions (per 100 people); Mortality rate, under-5 (per 1000 live births); Population density (people per sq. km of land area); Population growth (annual %); Population, total; Surface area (sq. km); Total debt service (% of exports of goods, services and primary income); Urban population growth (annual %).

2.3.2 Education Statistics

The World Bank EdStats Query holds approximately 2500 internationally comparable education indicators for contact purpose, evolution, conclusion, literacy, educators, population, and money spent by countries. The indicators elaborate the education sequence from preprimary to tertiary education. The inquiry placed on huge repository of datasets provides data from international learning assessments, equity data from domestic analysis, and protuberance data till year 2050. The attributes selected for this study are as follows: Government expenditure on teaching and learning as % of GDP (%); Gross registration ratio, primary, female (%); Gross registration ratio, primary, gender equality index; Net registration rate, preprimary, female (%); Net registration rate, primary, female (%); Out-of-school children of primary school age, both sexes (number); Percentage of all students in tertiary education registered in ISCED 6 and 7, both sexes (%); Primary achievement rate, both sexes (%); Ratio of girls to boys in primary and secondary education (%); School registration, primary (% gross); School registration, secondary (% gross).

2.3.3 Health, Nutrition and Population Statistics

Data about Key health, Nutrition and population statistics gathered from a variety of international sources has been utilized for the study. The indicators used are as follows: Adolescent fertility rate (births per 1000 women ages 15-19); Health expenditure per capita (current US$); Health expenditure, total (% of GDP); Health expenditure, total (current US$); Immunization, DPT, measles, polio; Improved cleanliness facilities (% of population with access); Improved cleanliness facilities, urban (% of urban population with access); Improved water resource (% of population with access); Improved water resource, urban (% of urban population with access); occurrence of tuberculosis (per 100,000 people); Out-of-pocket health spending (% of private expenditure on health); Out-of-pocket health spending (% of total spending on health); occurrence of anemia amid children (% of children under 5); occurrence of anemia amid nonpregnant women (% of women ages 15-49); occurrence of anemia amid pregnant women (%); occurrence of tuberculosis (per 100,000 population); Tuberculosis case recognition rate (all forms); Tuberculosis death rate (per 100,000 people).

2.4 Data Preprocessing and Feature Reduction

Data for the analysis is approximately 90% complete. The missing values have been filled up with estimated values using linear regression model.

Pearson's correlation coefficient has been calculated to find out the statistical relationships between two or more variables of health, climate change, education and other development related indicators so as to perform feature reduction. In positively correlated variables, increase or decrease in the value of one variable, causes the value of other variable also to increase or decrease accordingly. In negatively correlated variables, the value of one variable decreases as the value of the other increases and vice versa. Out of the 57 attributes, there are many that are strongly correlated and there are certain findings that are quite unexpected and are discussed in Results and Discussions section. Further only following 11 attributes, as given in Table 1, have been selected based on correlation coefficient values, for anticipation of GDP corresponding to HI and UMI groups using Artificial Neural Networks.

Table 1 List of 11 selected indicators for forecasting GDP (current US$)

Time	Mobile cellular subscriptions (per 100 people)
Adolescent fertility rate (births per 1000 women ages 15–19}	Mortality rate, under-5 (per 1000 live births)
Agriculture, value added	Population, total
CO_2 emissions (metric tons per capita)	Ratio of girls to boys in primary and secondary education (%)
Life expectancy at birth, total in years	Health expenditure, total (current US$)
	Improved sanitation facilities (% of population with access)

3 About Artificial Neural Network

An Artificial Neural Network is a numerical representation designed with motivation from the composition of biological neural networks. A neural network comprises of an interrelated collection of artificial neurons, and the neurons process information using a connectionist method to calculation [13, 14].

In this paper, ANN has been implemented using Matlab 7.0 software (matrix laboratory). A two layer MLP Back Propagation network with default settings has been applied for the training and testing of artificial neural network. In hidden layer of the back propagation network, tangent-sigmoid transfer function is employed. A pure linear transfer function is utilized in the output layer. The Back propagation learning algorithms, namely trainlm has been used for prediction of percentage of GDP. The input dataset comprises of attributes selected based on correlation coefficient, mentioned in Table 1. The output data corresponds to GDP of the country under consideration. A snapshot of dataset is illustrated in Table 2. Columns 1 to 12 of this table are used as input and column 13 corresponds to target. The range of inputs and outputs have been adjusted and hence lies in the range [−1,1]. A predefined function premnmx () provided by Matlab has been used for the purpose.

4 Results and Discussions

The Correlation coefficients between all 57 variables under the categories have been calculated and the variables having positive correlation of more than 0.9 have been selected. Since a modification in the value of one variable will forecast a modification in the similar way in the second variable, we have selected a subset of variables with 11 features to be used for forecasting GDP in case of UMI and HI. There are certain correlations that are implied but there are certain findings which are quite novel, unusual and interesting.

Corresponding to Upper Middle Income group of countries, it has been found that there exist very strong correlation coefficient of more than 0.9 between External debt stocks, Health expenditure and Foreign direct investment, net inflows. Also, GNI per capita and Health expenditure per capita (current US$) are strongly correlated with correlation coefficient more than 0.9.

Whereas in case of High Income group, the Health Expenditure is strongly correlated with CO_2 emissions, Electric Power consumption and Energy use. For UMI countries, the Health expenditure is weakly correlated (correlation coefficient = 0.4) with features like carbon dioxide emissions (metric tons per capita) and Electric power consumption.

Table 2 Sample of dataset for anticipation of GDP

Country name	Time	Adolescent fertility rate (births per 1000 women ages 15–19)	Agriculture, value added (% of GDP)	CO_2 emissions (metric tons per capita)	Life expectancy at birth, total (years)	Mobile cellular subscriptions (per 100 people)	Mortality rate, under-5 (per 1000 live births)	Population, total	Ratio of girls to boys in primary and secondary education (%)	Health expenditure, total (current US$)	Improved sanitation facilities (% of population with access)	GDP (current US$)
France	2000	10.39	2.34	5.98	79.06	49.06	5.4	60,911,057	99.744	1.34E+11	100	1.37E +12
France	2002	10.77	2.23	6.13	79.26	64.26	5.1	61,803,229	99.904	1.53E+11	100	1.50E +12
Germany	2012	3.80	0.87	8.70	80.89	111.59	4	80,425,823	96.119	3.86E+11	100	3.53E +12
Germany	2008	8.58	0.91	9.53	79.74	126.56	4.4	82,110,097	98.234	3.88E+11	100	3.75E +12
Japan	1998	4.68	1.71	9.17	80.50	37.77	5	1.26E+08	100.732	2.81E+11	100	3.91E +12
Japan	1999	4.96	1.68	9.46	80.57	45.30	4.7	1.27E+08	100.672	3.30E+11	100	4.43E +12
United Kingdom	2011	25.78	0.64	7.09	80.95	123.60	5	63,258,918	100.454	2.27E+11	100	2.59E +12
United Kingdom	2012	25.76	0.68	7.00	80.90	124.76	4.8	63,700,300	100.049	2.28E+11	100	2.61E +12
United States	1997	51.33	1.37	19.71	76.43	20.14	8.9	2.73E+08	101	1.12E+12	99.7	8.61E +12
United States	1998	49.70	1.24	19.62	76.58	24.89	8.7	2.76E+08	103.457	1.19E+12	99.7	9.09E +12
United States	1999	48.08	1.17	19.75	76.58	30.58	8.6	2.79E+08	101	1.26E+12	99.7	9.66E +12

Hence after selecting the important features based on correlation coefficient, we have trained the ANN with two different datasets, for UMI and HI countries, so as to forecast GDP. For UMI countries, the performance measurements in terms of Mean Square Error between actual and forecasted GDP stands at 0.00339 at 3 epochs, which is quite convincing, shown in Fig. 1. For HI countries, the performance measurement in terms of Mean Square Error between actual and forecasted GDP stands at 0.112 at 3 epochs, shown in Fig. 2.

Fig. 1 Result of training, testing and validating ANN for UMI countries data using learning function trainlm

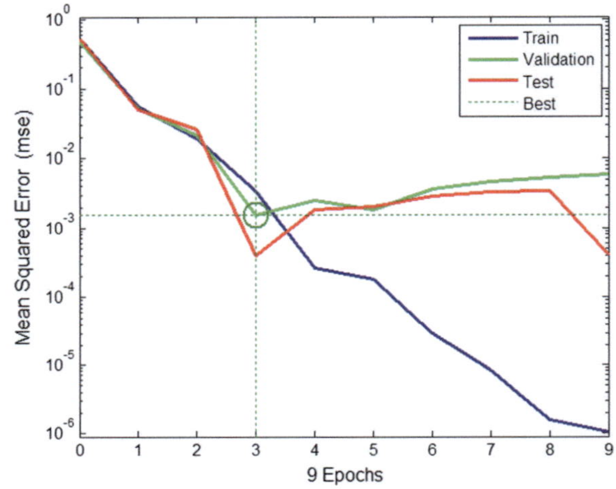

Fig. 2 Result of training, testing and validating ANN for HI countries Data using learning function trainlm

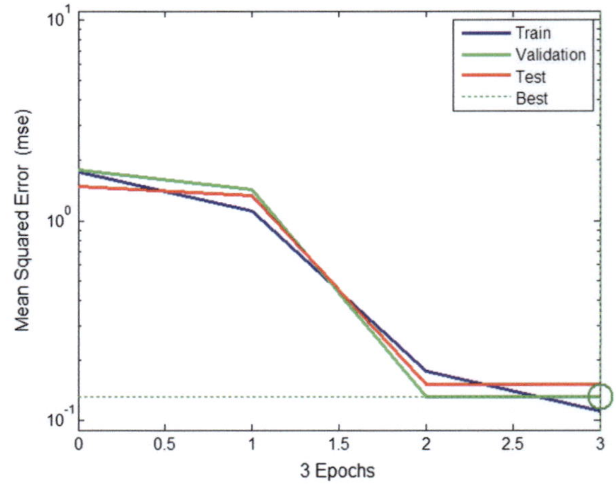

Table 3 Details of values of R corresponding to different datasets under consideration

S.no.	Datasets	R for UMI countries	R for HI countries
1	Training	0.99986	0.99997
2	Validation	0.99196	0.99879
3	Testing	0.99402	0.99906
4	All	0.99694	0.9995

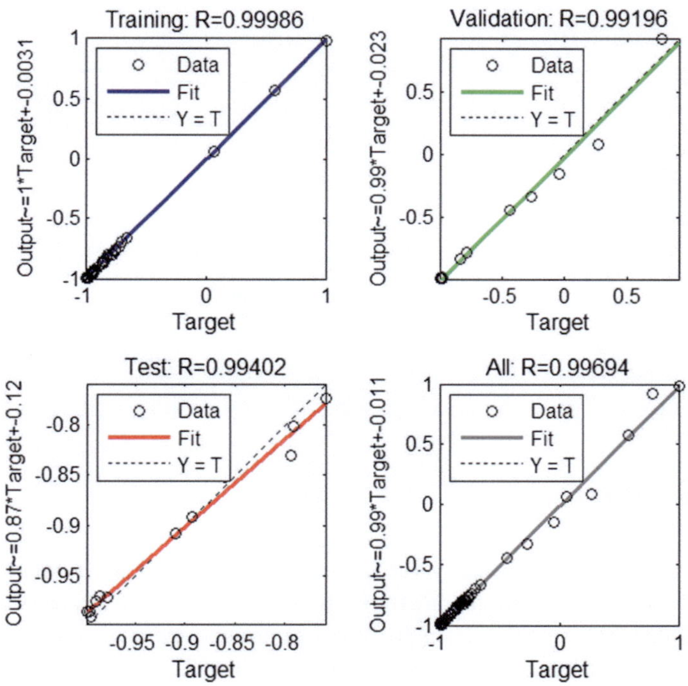

Fig. 3 Correlation analysis between Actual GDP and Predicted GDP for UMI countries

The correlation coefficient (R) representing the measure of linear relationship between actual GDP and the predicted GDP has been found as mentioned in Table 3 and shown in Figs. 3 and 4, respectively, for UMI and HI countries. A very strong correlation coefficient has been reported that advocates of the fact that the ANN with trainlm back propagation learning algorithm, transfer function, namely tangent-sigmoid applied in hidden layer of the back propagation network and pure linear transfer function utilized in the output layer is able to predict the GDP correctly.

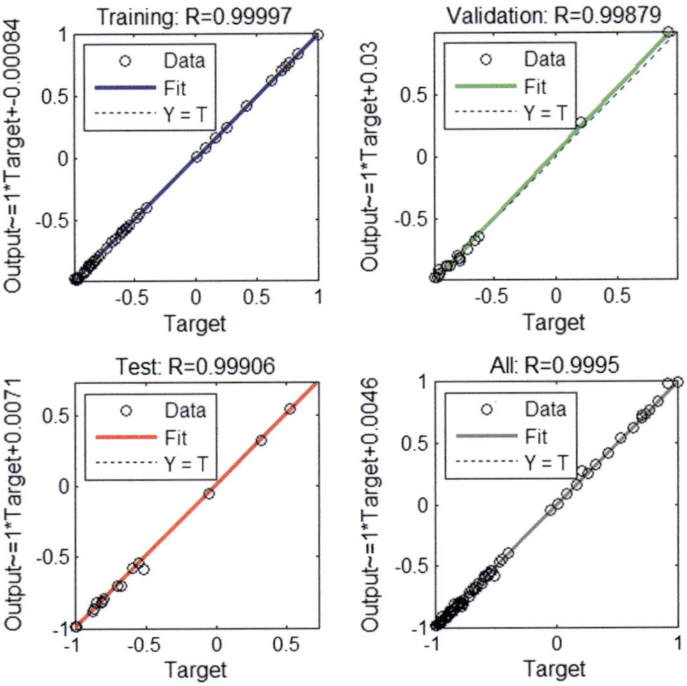

Fig. 4 Correlation analysis between Actual GDP and Predicted GDP for HI countries

5 Conclusion

Important correlations between various World Development Indicators have been found and the different correlations for UMI and HI countries have been discussed. It is concluded that ANN has been proved to be providing convincing results and is very suitable for anticipating Gross Domestic Product. Using the input parameters describing Education, Health, Climate change indicators, the ANN has been trained to predict the GDP for Upper Middle Income and High Income group countries. This research has noticeably demonstrated that application of Soft Computing techniques can facilitate in providing progressive details for forecast of GDP.

This paper differs from all of the earlier studies explained in Introduction section, as it is based not only on economical indicators but also on Education statistics and Health, Nutrition and population statistics. Also the anticipation of GDP is not done just for a single country but on UMI and UI segment of economies. In this paper, selection of a small subset of features has been done on the basis of strong correlations between more than 50 indicators. Also, a few unusual correlations have been observed that are discussed in Results section.

The study is further to be extended for finding correlations between different development indicators for remaining categories of economies.

Acknowledgments The author would like to express deepest sense of gratitude to Prof. (Dr.) R. K. Datta, a renowned personality in the field of Information Sciences and Meteorology, currently Director, Mohyal Educational Research Institute of Technology, for his encouragement, guidance and mentoring. Without his support, it would not have been possible to take up research in this challenging field.

References

1. Economic Development and the Risk of Global Climate Change.: Available from: www.worldbank.org/depweb/beyond/beyondbw/begbw_14.pd
2. Smith, J.B.: Vulnerability to climate change and reasons for concern: a synthesis. In: McCarthy J.J., et al. Climate Change 2001: Impacts, Adaptation and Vulnerability. Contribution of Working Group II to the Third Assessment Report of the Intergovernmental Panel on Climate Change. Cambridge University Press, Cambridge, UK, and New York, N.Y. U.S.A. (2001)
3. Bhargava, A., Jamison, D.T., Lau, l., Murray, C.J.L.: Modeling the effects of health on economic growth, gpe discussion paper series: no. 33 evidence and information for policy world health organization
4. David, E., Bloom, D.E., Canning, D., Sevilla, J.: The Effect of Health on Economic Growth: Theory and Evidence, NBER Working Paper No. 8587, Issued in November 2001
5. From: Education at a Glance:. Highlights Access the complete publication at: http://dx.doi.org/10.1787/eag_highlights-2012-en (2012)
6. Hanushek, E.A., Jamison, D.T., Jamison, E.A., Woessmann, L.: Education and Economic Growth, Education and Economic growth, Education Next, vol. 8, no. 2. Spring (2008)
7. Swanson, N.R., White, H.A.: RA model selection approach to real time macroeconomic forecasting using linear models and artificial neural networks. Rev. Econ. Stat. **79**, 540–550 (1997)
8. Kohzadi, N., Boyd, M.S., Kaastra, I., Kermanshahi, B.S., Scuse, D.: Neural networks for forecasting: an introduction. Canadian J. Agr. Economics/Revue canadienne d'agroeconomie **43**(3), 463–474 (1995)
9. Li, Y.: Macroeconomics modeling on UK GDP growth by neural computing, technical report, CSC-95009, (1995)
10. Tkacz, G., Hu, S.: Forecasting GDP growth using artificial neural networks, Working paper 1999-3/ Bank of Canada, pp. 1–24, (1999)
11. Junoh, M.Z.H.M.: Predicting GDP growth in Malaysia using knowledge-based economy indicators: a comparison between neural network and econometric approaches. Sunway Acad. J. **1**, 39–50 (2004)
12. Zhang, G.P.: Time series forecasting using a hybrid ARIMA and neural network model. Neurocomputing **50**, 159–175 (2003)
13. Sivanandam S.N., Sumathi S., Deepa S.N.: Introduction to Neural Networks using Matlab. Tata McGraw Hill Education Private Ltd., (2009)
14. Kosko B.: Neural Networks and Fuzzy Systems. Prentice Hall of India Ltd., (2005)

An Efficacious Matching of Finger Knuckle Print Images Using Gabor Feature

Nivedita Bhattacharya, Deepak Kumar Dewangan and Kranti Kumar Dewangan

Abstract An efficacious matching of 240 Finger knuckle print images using Gabor feature by extracting position and weight with a record of 5362 values for each image. In this approach the original image is compared with the other images captured at different angles. According to the obtained records a difference is calculated between the two images which results in how much disparity subsist between the two images. The comparison is made apparent by generating a 3-D graph which shows the best match of the original image. The data is taken from Hong-Kong Polytechnic University which contains 7920 FKP images of 660 different individuals.

Keywords Finger knuckle print · Gabor · Matching · Position · Weight

1 Introduction

On the basis of physiological or behavioral characteristics "Biometrics" is defined as an automated method of identifying or recognizing an individual uniquely. Under physiological characteristics some of the biometrics is face, palm veins, palm print, fingerprint, hand geometry, iris, and DNA. Under behavioral characteristics some of the biometrics is voice, signature, typing pattern and gaits. The traditional approach of identifying the human beings dates back to 1870s. The Bertillon's system was used for measuring the body parts such as arm and foot length measurements and skull diameter of the prisoners to identify them in the USA until 1920s [1]. A quantitative approach for identification through facial measurements

N. Bhattacharya (✉) · D.K. Dewangan · K.K. Dewangan
Computer Science & Engineering, Raipur, C.G, India
e-mail: niveditabhattacharya.6580@gmail.com

D.K. Dewangan
e-mail: deepakdewangan27@gmail.com

K.K. Dewangan
e-mail: kranti.d123@gmail.com

A.K. Saini et al. (eds.), *ICT Based Innovations*, Advances in Intelligent Systems and Computing 653, https://doi.org/10.1007/978-981-10-6602-3_15

and fingerprint was proposed by William Herschel, Henry Faulds and Sir Francis Galton [1, 2]. In the early 1960s voice and fingerprint technologies were used to provide high-security for access control, financial transactions and personal locks [1, 3]. In 1970s the hand geometry systems came into existence [1, 4]. In 1980s ejaculated the signature verification [5, 6] and retinal systems [7, 8]. Iris recognition systems came in 1990s [9, 10]. Biometric authentication is performing a crucial role in access control and public security. The most often used biometric traits are signature, iris, fingerprint, gaits, face etc. Some recent advances have been made in emerging biometrics. In the Holocene years biometric technologies based on heart (electrocardiogram) and brain (electroencephalogram) signals have emerged [11].

Compared to formal biometrics this technology has an advantage of being more impostor repellent. Investigation of biometric traits nowadays is highly done by researchers so as to incarnate a system for identifying an individual uniquely.

A newly biometric technology which has fascinated the attention of researchers is the finger knuckle print [11]. Finger knuckle refers to the inborn patterns formed at the conjunct at the back surface of the finger. There are some strong real applications of finger knuckle print which makes it highly unique and they are: (1) Since people hold the objects with the inner side of the hands so it cannot be eroded easily. (2) Unlike fingerprint there is no smudge of felonious investigation associated with it, thus the user acceptance rate is higher. (3) Finger knuckle print is rarely left by people on credential surface due to which the possibility of losing private data is less. Thus finger knuckle print is deliberated as one of the most auspicious biometric techniques for identifying a person uniquely [11, 12]. Alike different biometric techniques matching and extracting the feature plays a crucial role in finger knuckle print recognition.

2 Literature Review

In 2009, L. Zhang, D.Zhang et al. extracted information about local orientation using two-dimensional Gabor filter and similarity was measured between two competitive coding maps using angular distance [13].

In 2011, Wankau YANG, Changyin SUN et al. proposed a combined approach of Gabor feature and MMDA to identify finger knuckle print. Firstly Gabor feature is calculated then after MMDA transformation is done and finally the Gabor feature representation is projected by the projection matrix and classified. The combined approach resulted with a method having good performance [14].

Tao Kong et al. in 2014 proposed a hierarchical classification method for recognizing finger knuckle print. The method consists of two phases in the first phase basic feature is conducted and in the second phase minor is conducted which results in balance between verification speed and accuracy [12].

In 2010, L.Zhang et al. presented an online personal authentication using FKP (Finger knuckle print) that has different line features. Gabor filters is used for extracting the orientation and magnitude. In this a image acquisition device and the associated data processing algorithm is developed including a cost-effective FKP system which results in high accuracy, small-size, high speed and cost-effective [15].

Meng Yang, Lei Zhang proposed a Gabor feature based SRC (sparse representation based classification) dodge and proposed a comrade Gabor occlusion dictionary calculating algorithm to haft the occluded face images. One important advantage of Gabor sparse representation based classification is its compact occlusion dictionary which contains fewer atoms as compared to the original scheme which effectively reduces the computational cost of sparse coding. This method results in better performance than SRC [16].

In 2011, ZHU Le-qing proposed a finger knuckle print recognition algorithm based on Surf algorithm. Firstly to align the images a coordinate system is defined on the basis of local convex direction map of FKP and for feature extraction a region of interest (ROI) is cropped. Secondly, fast Hessian Detector is used for extracting the key points. In this paper RANSAC based matching scheme and SURF based extraction of features is introduced. This method results in with high performance in terms of efficacy and delicacy. Also the method is invariant to scaling, rotation and translation of image. Surf is used for improving the time performance in SIFT.

3 Methodology

3.1 Introduction

Set of images $B_j = [p_{j,1}, p_{j,2}, p_{j,3} \ldots p_{j,nj}] \in S_j^{m*n}$ of the jth object class, where $p_{j,i}$ ($i = 1,2,\ldots\ldots,n_j$) is an m-dimensional vector resilient by the ith sample of the jth class. For a test sample $x_0 \in S^m$ from this class, naturally x_0 can be well relative by the linear mixture of the samples within B_j, i.e. $x_0 = \Sigma_{i=1}\, \alpha_{j,i}\, p_{j,i} = B_j\, \alpha_j$, where $\alpha_j = [\alpha_{j,1}, \alpha_{j,2}, \ldots\ldots\, \alpha_{j,nj}]^R \in S_j^n$ are the coefficients. Estimate we have N object classes and let $B = [B_1, B_2, \ldots\ldots, B_N]$ be the conjugate of the n training samples from all the N classes, where $n = n_1 + n_2 + \ldots\ldots + n_N$, then the linear representation of x_0 can be written in stint of all training samples as $x_0 = B\alpha$, where $\alpha = [\alpha_1; \ldots; \alpha_j; \ldots \alpha_N] = [0, \ldots, 0, \alpha_{j,1}, \alpha_{j,2}, \ldots, \alpha_{j,nj}, 0, \ldots, 0]^R$.

3.2 Algorithm

(i) Input image
(ii) Convert it into rgb2gray
(iii) Calculating Matrix using the Gabor filter

(iv) Implementing Gabor filters
(v) Down sampling the images in context of rows and columns
(vi) Calculation of Filtration of image by calling the matrix resulted in step-(iii)
(vii) Extracting the feature in terms of position and weight
(viii) End

In this paper, Gabor feature is used for extracting the position and weight of finger knuckle print with a record of 5362 values for each image. 240 finger knuckle images were tested with this method and each image generated a record of 5362 values. For each image a difference (d) is calculated and then each FKP image is compared with the other images. The comparison is done on the basis of one original image of one finger of an individual with the same images of that finger at different angle. The comparison is made visible with the 3-D graphs plotted for the images. The comparison of FKP images results in the best match of the original image.

4 Result

When the images are applied on the above algorithm it results with position and weight with a record of 5362 values for each image. Then a difference (d) is calculated for each image which shows how much distinct is the other image from the original. A total of 240 images were tested in which there is a set of 12 images for one finger from which one is original and other images are at different angles. The comparison is shown with the 3-D graphs below: (Figs. 1, 2, 3, 4, 6, 7, 9, 10 and 11)

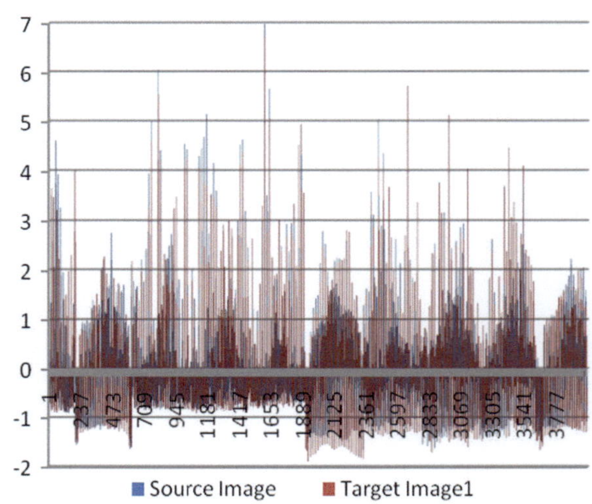

Fig. 1 Comparison between the Source Image and Target Image1, with difference $d = 11.93$

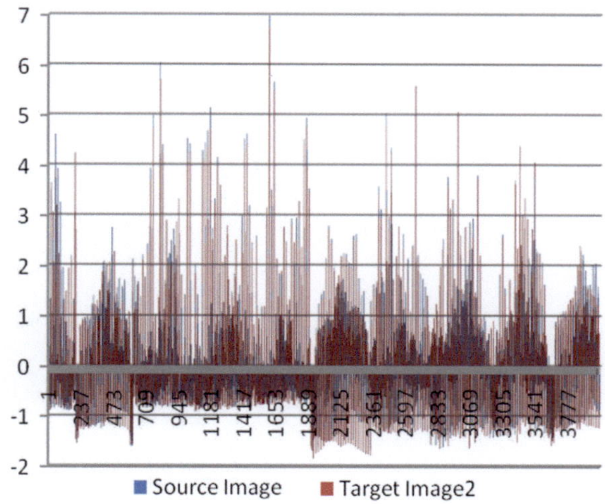

Fig. 2 Comparison between the Source Image and Target Image2, with difference $d = 12.10$

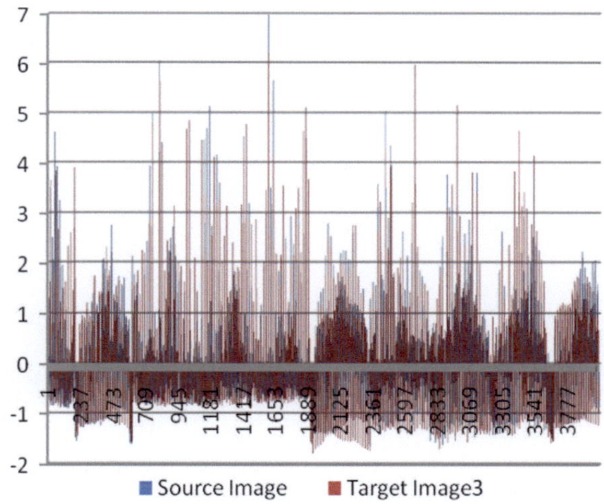

Fig. 3 Comparison between the Source Image and Target Image3, with difference $d = 7.82$

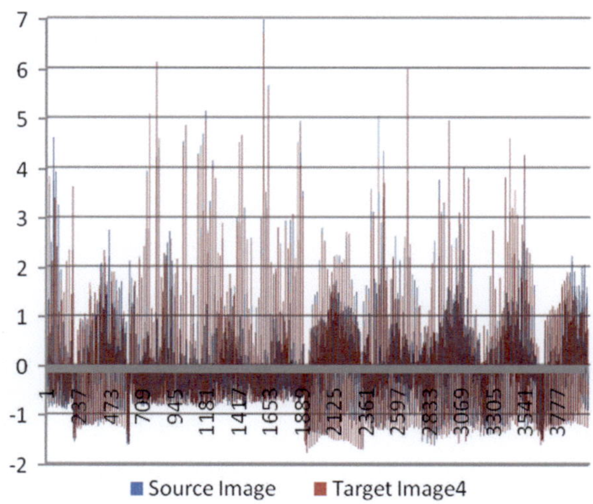

Fig. 4 Comparison between the Source Image and Target Image4, with difference $d = 0.562$

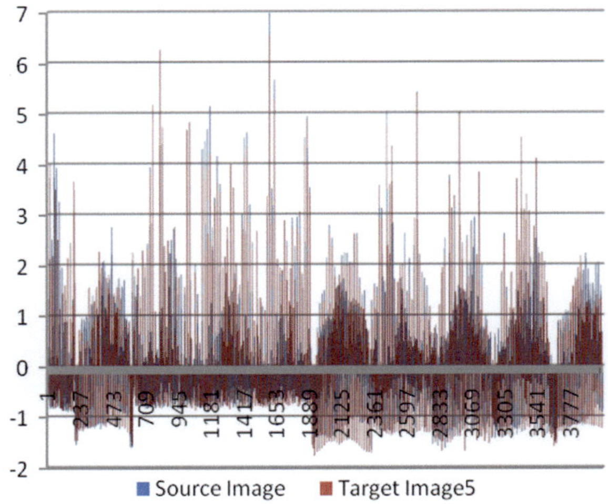

Fig. 5 Comparison between the Source Image and Target Image5, with difference $d = 14.88$

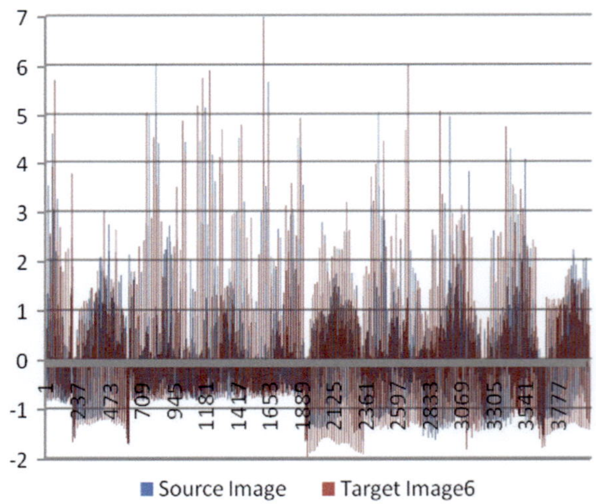

Fig. 6 Comparison between the Source Image and Target Image6, with difference $d = 14.47$

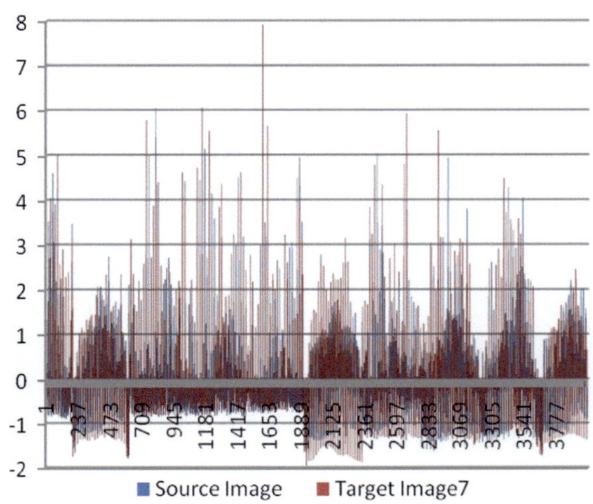

Fig. 7 Comparison between the Source Image and Target Image7, with difference $d = 47.07$

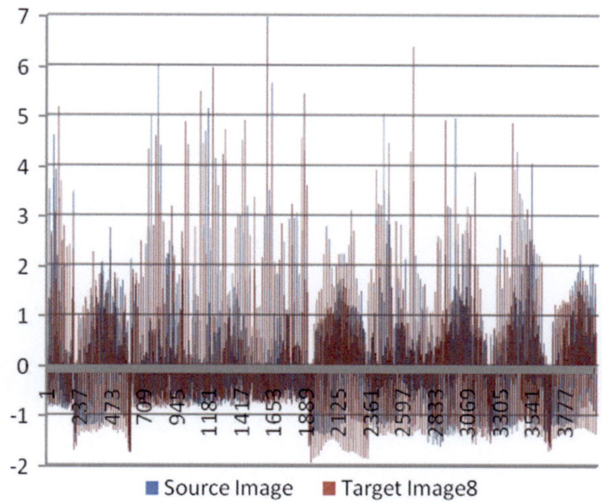

Fig. 8 Comparison between the Source Image and Target Image8, with difference $d = 18.44$

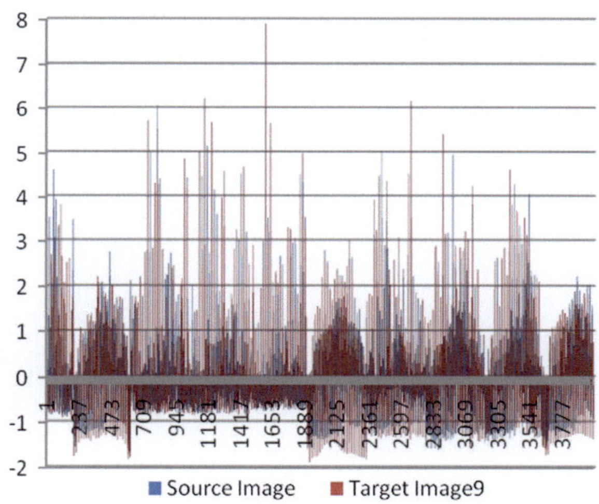

Fig. 9 Comparison between the Source Image and Target Image9, with difference $d = 24.23$

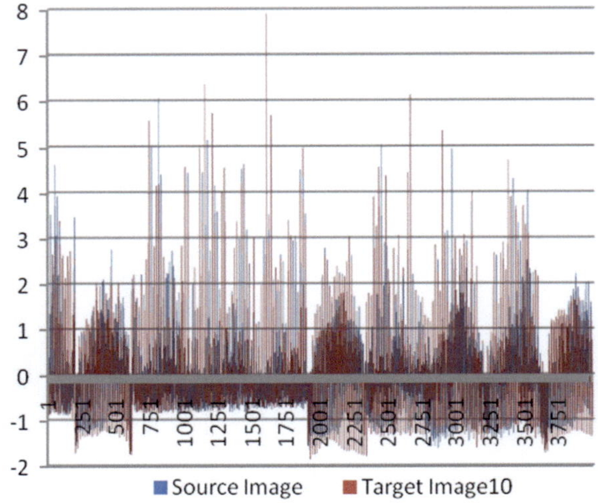

Fig. 10 Comparison between the Source Image and Target Image10, with difference $d = 17.36$

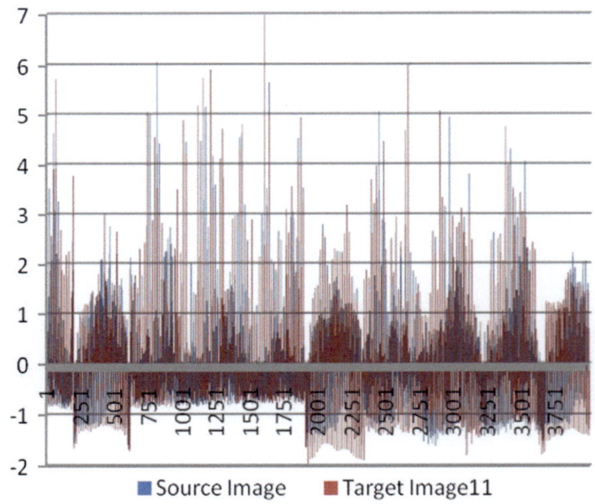

Fig. 11 Comparison between the Source Image and Target Image11, with difference $d = 14.47$

5 Data Set

The database is collected from Hong Kong polytechnic University containing 7920 sample images of FKP from 660 individuals. Each sample contains 12 images of each four fingers (left index, left middle, right index, right middle). Also it contains 7092 sample images of ROI. Samples were collected in two separate sessions (Figs. 12 and 13).

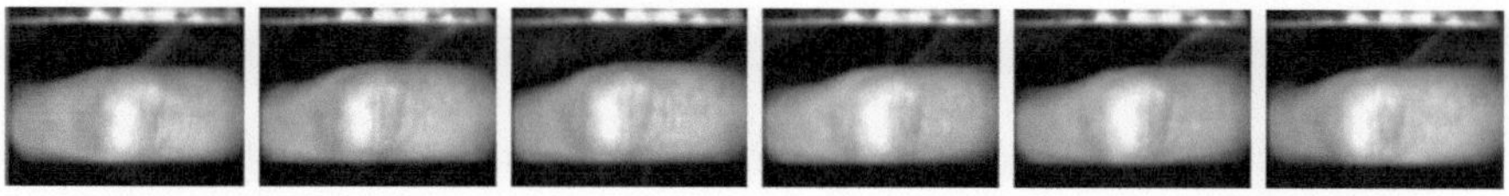

Fig. 12 FKP Database

Fig. 13 ROI Database

6 Conclusion

In this paper Gabor feature is used for comparing the FKP images and resulting in best match. A record of 5632 values of position and weight were obtained for each image. Then a difference was calculated which shows how much distinct the target image is from the original image.

The maximum difference between the source image and the target images is $d = 47.07$ which defines the worst match and the minimum difference between the source image and target images is $d = 0.562$ which defines the best match. Figures 5 and 8 shows the best match and worst match respectively. 240 images were tested in this approach and compared using the graph. Here sample of 12 tested images is been displayed. This approach has advantages like accurate, flexible and effective. Assimilate with other subsisting finger rear surface based systems, the propound one has grace of accuracy, flexible and efficient.

7 Future Scope and Limitations

Implementation of Finger Knuckle with the integration of upcoming future algorithm will surely enhance the security mechanism. Till now Finger Knuckle and recent security technique is working offline, in future it can be operated with Cloud computing/Internet also. This test is applied on static images; it can also be applied on dynamic images or videos for recognizing Finger Knuckle Print.

The limitations of Finger Knuckle Print can be hairs on the knuckle and cut or wounds on the knuckle.

References

1. Wayman, J., Jain, A., Maltoni, D., Maio, D.: An Introduction to Biometric Authentication Systems. Biometric Systems technology, Design and Performance Evaluation ISBN 978-1-85233-596-0 Springer (2005)
2. Faulds, H.: On the skin furrows of the hand. Nat. 22, 605 (1880)
3. Trauring, M.: On the automatic comparison of finger-ridge patterns. Hughes Laboratory Research Report No. 190 (1961)
4. Zunkel, R.: Hand Geometry based verifications. Biometrics: Personal Identification in Networked Society. Kluwer Academic Press (1999)
5. Samples, J.R., Hill, R.V.: Use of infrared fundus reflection for an identification device. **98**(5), 636–640, (1984)
6. Hill, R.H.: Retina Identifiocation, Biometrics: Personal Identification in Networked Society. Kluwer Academic Press (1999)
7. Crane, H.D., Ostrem, J.S.: Automatic Signature Verification using a three axis force-sensitive pen. IEEE transactions on Systems, Man and Cybernetics, SMC-**13**(3), 329–337 (1983)
8. Nalwa, V.S.: Automatic Online signature Verification. **85**(2), 215–239 IEEE (1997)
9. Wildes, R.P.: Iris Recognition: an emerging biometric technology, IEEE. **85**(9), 1348–1364 (1997)
10. Jain, A., Bolle, R., Pankati, S.: Introduction to Biometrics, Biometrics: Personal Identification in Networked Society. Kluwer Academic Press (1999)
11. Bhattacharya, N., Dewangan, D.K.: Implementation and Assessment for identification of Finger knuckle using Scale Invariant Feature Transform technique. Int. J. Appl. Eng. Res. ISSN **10**(44),0973–4562 (2015)
12. Kong, T., Yang, G., Yang, L.: A Hierarchical classification method for finger knuckle print recognition. EURASIP. J. Adv. Sign. Proces. A Springer open journal
13. Zhang, L., Zhang, L., Zhang, D.: Finger knuckle print verification based on band- limited phase-only correlation. 13th International Conference on Computer analysis of images and patterns, (Munster) Germany 2009
14. Yang, W., Sun, C., Wang, Z.: Finger knuckle print recognition using Gabor feature and MMDA. Front. Electr. Electron. Eng. China. **6**(2), 374–380 (2011) doi: 10.1007/s11460-011-0141-3
15. Zhang, L., Zhang, L., Zhang, D., Zhu, H.: Online finger knuckle print verification for personal authentication. Pattern recognition **43,** 2560–2571. Elsevier (2010)
16. Yang, M., Zhang, L.: Gabor feature based sparse representation for face recognition with gabor occlusion dictionary. Biometric Research Center, Dept. of Computing, The Hong Kong Polytechnic University, Hong Kong
17. Bhattacharya, N., Dewangan, D.K.: Fusion technique for Finger Knuckle Print recognition. International Conference on Electrical, Electronics, Signals, Communication and Optimization(EESCO) 978-1-4799-7678-2/15 IEEE, (2015)

An Ensemble-Based Decision Support System for the Students' Academic Performance Prediction

Mrinal Pandey and S. Taruna

Abstract Students' academic performance analysis and prediction activities enable the educators to prevent their students from failure by providing necessary support and counseling. With this necessity, an analysis-based Decision Support System (DSS) is proposed for the prediction of the students' performance using ensemble methods. Nowadays ensemble methods becoming more popular in classification and prediction. In ensemble-based methods, a set of classifiers are used to construct a single composite classifier and the decisions of the individual classifiers are combined using some combination rules to classify the new instances. We proposed an ensemble-based Decision Support System (DSS) for the prediction of students' performance particularly in the engineering discipline.

Keywords Classification · Prediction · Performance · Accuracy · Ensemble · Decision support system

1 Introduction

An accurate prediction of students' academic performance at early stages of the degree program helps to identify weak students and enables the management to take the corrective actions to prevent their students from failure. In this research, three classifiers namely Naïve Bayes, K-Nearest Neighbor (IBK), and Decision Tree [1] are combined and a novel hybrid ensemble of classifiers is proposed to predict the performance of the students at different stages (level) during their engineering graduation program.

M. Pandey (✉) · S. Taruna
Banasthali University, Rajasthan Vanasthali, India
e-mail: mrinalpandey14@gmail.com

S. Taruna
e-mail: staruna71@yahoo.com

A.K. Saini et al. (eds.), *ICT Based Innovations*, Advances in Intelligent Systems and Computing 653, https://doi.org/10.1007/978-981-10-6602-3_16

2 Literature Review

A number of researches related to the ensemble techniques have been carried out by many researchers in the area of educational data mining. A comparative study of ensemble techniques, namely Boosting, Bagging, Random Forest, and Rotation Forest was conducted for students' performance prediction [2]. In [3] Adaboost ensemble with the genetic algorithm was proposed to predict the performance of the students in early stages so that the risk of failure can be reduced by providing appropriate advice to the students who are at high risk. Authors of [4] combined the HNB and Decision stumps (DS) algorithms using voting technique and compared the proposed algorithm with a number of classifiers for two-class predictions, three-class predictions and four-class predictions particularly for student performance prediction. Authors of [5] conducted a study of the local voting scheme of classifiers using three weak learning algorithms, namely OneR, Decision Stump, and Naive Bayes. A hybrid decision support tool was developed particularly for the Hellenic Open University to identify drop-out prone students [6]. In this study Naive Bayes, C4.5, and BP algorithm were combined using a stacking variant methodology.

3 Ensemble Methods

Ensemble method combines the predication of various base classifiers and gives the final prediction. The ensemble model combines the set of classifiers to create a single composite model, which provide better accuracy. Ensemble methods can be termed as a committee, classifier fusion, combination, or aggregation, etc. Research shows that the prediction from a composite model gives the best results as compared to a single model prediction. A number of methods have been proposed for the ensemble construction in [7] such as an ensemble can be constructed by using different subsets of training data with a single learning method, for e.g., Bagging [8], Boosting [9], Random Forest [10], etc. Other types of ensemble can be constructed by using different training parameters with a single training method, for e.g., neural network, it can also be constructed using a combination of different learning methods, for e.g., stacking and voting.

4 Proposed Method

In this research, one weak classifier namely Naïve Bayes and two strong classifiers namely K-Nearest Neighbor (IBK) and Decision Tree (J48) were selected and proposed a hybrid ensemble of classifier (JIN) using the voting technique. The

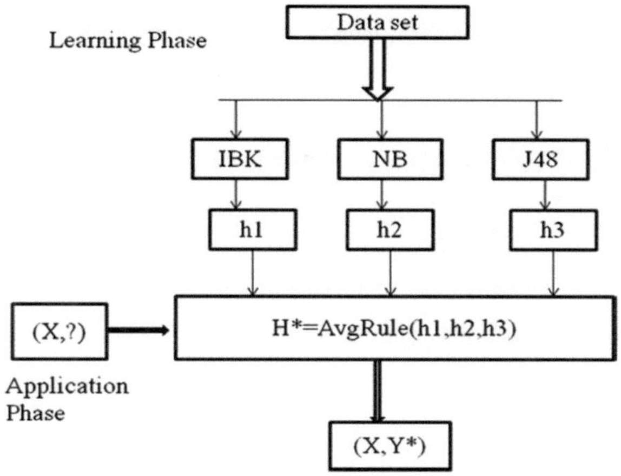

Fig. 1 Ensemble construction (Proposed Model)

Naïve Bayes classifier and the Decision Tree classifiers were selected for better visual representation. Additionally, Naive Bayes is a linear classifier and less likely to suffer from over fitting problem [5], while K -Nearest Neighbor (IBK) performs better for large datasets and complements the DT algorithm. All three classifiers are combined using average majority voting combining rule. In this experiment h1, h2, and h3 are the hypothesis obtained from each individual classifier namely J48, IBK, and NB respectively. For each output class, a posteriori probability is generated by the individual classifier. The class represented by the maximum average value of a posteriori probabilities is selected to be the voting hypothesis (h*) for the final decision. Figure 1 shows the proposed model for our study and Fig. 2 presents proposed JIN algorithm.

5 Experimental Setup

The data for this study is collected from an engineering college from India. The dataset consists of 1000 instances, student's demographic information and academic performances from high school to the final year of the engineering course. The WEKA [11] toolkit is used for the analysis. A number of attribute selection methods have been applied such as gain ratio, info gain, and chi-square for attribute selection. Finally, the gain ratio measure is used for attribute selection. It was observed that the academic attributes have higher gain ratio values as compared to the demographic attributes. Therefore, only performance attributes were selected for this study. It is a multi-level prediction, therefore, the performance of the students' is predicted at different stages of the course and each time some new performance attribute is added to the previous attributes. Table 1 shows the list of attributes used at different levels for the multi-level prediction.

STEP 1:-Input the Training Dataset D with the class labels L={A,B,C,F} represents class J.

STEP 2:-Input the Learning Algorithm (J48, NB, IBK)

STEP 3:-Call a set Eensemble (J48, IBK NB) and receive the hypothesis (classifier) h_n.

STEP 4:- Add h1,h2,h3 to Ensemble E.

STEP 5:- Perform voting for – Given unlabeled instance x

STEP A:- Evaluate the Ensemble E = $\{h_1, h_2, h_3\}$

STEP B:- Obtain average vote received by each output class using average of voting combining rule.

$$V_j(x) = \frac{1}{N}\sum_{n=1}^{N} P_n$$

Where: P_n is the posterior probability of the respective classifier for each output class J for the corresponding new instances x.

STEP C:- Choose the class that receives the highest total vote as the final classification.

$$C_{final}(x) = argmax\, V_j(x)$$

STEP 6:- END

Fig. 2 Proposed Algorithm: (JIN: J48 IBK, Naïve Bayes)

Table 1 Combined list of attributes used at various levels

Beginning of First year (Level-1)	Consists of the academic performance of the Student's grades in standard 10th, standard 12th, Name of High School board and Name of Senior Secondary board
Beginning of Second year (Level-2)	Consists of the attributes of the previous level as well as the performance of I year (first and second semester performances) in terms of grades
Beginning of Third year (Level-3)	Consists of the attributes of the previous level, the performance of the II year (third semester and fourth semester performances) and Backlogs
Middle of 4th years (Level-4)	Consists the attributes of the previous level, aggregate grade till third year (performances up to sixth semester) and the aggregate grade till the middle of the fourth year (performances up to seventh semester) and Backlogs

6 Empirical Results and Comparisons

Initially three classifiers namely IBK, NB, and J48 are used to construct the ensemble model using the voting methodology. A cross folds 10 validation is used to evaluate the prediction accuracy. The developed JIN model is compared with the individual classifiers on the basis of t-test with $p < 0.05$. Table 2 depicts the results of this comparison in terms of accuracy percentage. It is clear from the Table 2 that

Table 2 Comparison of individual classifiers with JIN in terms of accuracy percentage

Classifier	Voting (IBK + J48 + NB)	NB	IBK	J48
Beginning of First year (Level-1)	61.46	60.38*	61.04	61.49
Beginning of Second year (Level-2)	72.73	70.62*	72.96	72.45
Beginning of 3 years (Level-3)	76.9	76.46	75.3*	75.68
Middle of 4 years (Level-4)	85.82	84.52	85.48	86.38

* Indicates significantly better results of the proposed JIN classifier from respective classifiers

the accuracy of the prediction is highest at level 4 and lowest at level 1 for all the classifiers. It can also be observed that the proposed algorithm is more accurate than NB classifier at level 1 and level 2. It is also more accurate from IBK at level 3, while there is no significant difference at level 4 for any of the classifier. The proposed classifier is also compared with other well-known homogenous ensemble classifiers such as bagging and boosting variant of NB, IBK and J48 classifier for t-test with $p < 0.05$. The result of these comparisons with the JIN model is depicted in Table 3.

It can be observed from the Table 3 that boosting using NB as a base classifier is significantly less accurate in level 1 and level 2, respectively, while J48 boosting and IBK boosting is less accurate in level 3 than the proposed algorithm. Moreover, the results of bagging and boosting shows that NB as a base classifier is significantly less accurate than the proposed algorithm in level 1 and level 2. It is also noticeable that the prediction accuracy increases at each level, i.e., when the number of academic performance attributes increases than the prediction accuracy also increases.

7 Conclusion and Future Work

The notion of heterogeneous ensemble classifiers is used for the students' academic performance prediction. Three classifiers namely Naive Bayes, Decision Tree (J48) and K-Nearest Neighbor (IBK) were integrated using an average of voting combining rule and a hybrid classification model (JIN) is proposed for the forecasting the students' academic performance. The result shows that the proposed ensemble model is also better than homogenous classifiers. Thus the proposed ensemble-based decision support system is useful for predicting the student performance and enables the educators to take preventive actions for the success of the students and the growth of the organization. The hybrid model proposed in this research can also be used in various domains of data mining applications for determining the robustness of the proposed model as future work. Additionally, another ensemble technique such as stacking, mixture of experts, etc., can be explored for the development of decision support system.

Table 3 Comparison of Bagging & Boosting with JIN in terms of accuracy percentage

Classifier	Voting (IBK + J48 + NB)	Bagging NB	Bagging IBK	Bagging J48	Boosting IBK	Boosting NB	Boosting J48
Beginning of First year (Level-1)	61.46	60.39*	60.98	61.3	61.04	60.38*	61.49
Beginning of Second year (Level-2)	72.73	70.71*	72.6	72.82	72.96	70.62*	72.45
Beginning of 3 years (Level- 3)	76.9	76.64	75.49	75.62	75.3*	76.46	75.56*
Middle of 4 years (Level-4)	85.82	84.67	85.11	86.47	85.48	84.52	86.19

References

1. Quinlan, J.R.: C4.5 Programs for machine learning. Morgan Kaufmann, San Francisco (1993)
2. Pandey, M., Taruna, S.: A comparative study of ensemble methods for students' performance modelling. Int. J. Comput. Appl. (0975–8887) **103**(8), October 2014, pp. 26–32 (2014)
3. ElDen, A.S., Moustafa, M.A., Harb, H.M., Emara, A.H.: Adaboost Ensemble with Simple Genetic Algorithm for Student Prediction Model. Int. J. Comput. Sci. Inf. Technol. **5**(2), (2013)
4. Paris, I.H.M., Affendey, L.S., Mustapha, N.: Improving academic performance prediction using voting technique in data mining. World. Acad. Sci. Eng. Tech. **38**, 2010-02-25. (2010)
5. Kotsiantis, S., Pintelas, P.: Local voting of weak classifiers, Int. J. Knowl. based Intell. Eng. Syst. 9 (2005) 239–248 239, IOS Press. (2005)
6. Kotsiantis, Pintelas.: A hybrid decision support tool—using ensemble of classifiers. ICEIS 2004—Artificial Intelligence and Decision Support. (2004)
7. Dietterich, T.G.: Ensemble methods in machine learning. In Kittler, J., Roli, F. (eds.) Multiple Classifier Systems, LNCS, vol. 1857, Springer, pp. 1–15. (2001)
8. Breiman, L.: Bagging predictors. Mach. Learn. **24**(2), 123–140 (1996)
9. Freund, Y., Schapire, R.: Experiments with a new boosting algorithm, Proceedings: ICML'96, 148–156
10. Breiman, L.: Random forests. Mach. Learn. **45**(1), 5–32 (2001)
11. Weka, University of Waikato, New Zealand, http://www.cs.waikato.ac.nz/ml/weka/
12. Quinlan, J.R.: Bagging, Boosting, and C4.5. In Proceedings of the Thirteenth National Conference on Artificial Intelligence, pp. 725–730 (1996)
13. Popular Ensemble Methods: D. and Maclin, R. An Empirical Study, Journal of Artificial Research **11**, 169–198 (1999)

Adding Big Value to Big Businesses: A Present State of the Art of Big Data, Frameworks and Algorithms

D. Radhika and D. Aruna Kumari

Abstract Data plays a pivotal role in business growth. In fact, data is considered to be an asset to organizations. This is more evident in the enterprises where the data is preserved and mined for discovering knowledge. The data with exponential growth and characterized by volume, velocity, and variety is termed as big data. Mining such voluminous data can give comprehensive business intelligence for making strategic decisions. The emergence of cloud computing technology, parallel processing power of servers, and the distributed programming frameworks like Hadoop with new programming paradigm "MapReduce" pave way for mining massive-scale data. Data mining domain is rich in algorithms that are used to mine data for discovering trends. The era of big data has arrived and mining such data is beyond the capability of conventional data mining techniques. The unprecedented exponential growth of data needs a platform for effective data analysis in real time with fast response. In this paper, we present an overview of big data, mechanisms or algorithms and environment or tools needed to execute them. The rationale behind this paper is that big data mining is the need of the hour in all sectors like finance, biology, healthcare, banking, insurance, and environmental research to name few. Review of various aspects of big data mining can help readers to gain know-how in the context of globalization, business collaborations where mining cross-organization data is essential. This paper also throws light into the relationship among big data, cloud computing technology, Hadoop, and Big data storage systems. In future, we intend to propose and implement algorithms for big data mining.

Keywords Big data · Big data mining · Distributed programming frameworks · Algorithms · Hadoop

D. Radhika (✉)
Computer Science Engineering, K L University, Guntur, AP, India
e-mail: radhikarajasekhar@yahoo.com

D. Aruna Kumari
Department ECM, K L University, Guntur, AP, India
e-mail: aruna_d@kluniversity.in

A.K. Saini et al. (eds.), *ICT Based Innovations*, Advances in Intelligent Systems and Computing 653, https://doi.org/10.1007/978-981-10-6602-3_17

1 Introduction

In recent years, data is growing rapidly in such a way that it became difficult to handle. Not only in size, the data has acquired other properties as well forming Big data. Big data has properties like volume, velocity, and variety. Recently more attributes were added to the list that includes variability, complexity, and value [1]. Description of these attributes is given in Sect. 2. Traditional approaches to mine Big data are inadequate as processing Big data needs huge computing resources in the distributed environment. A new programming paradigm known as MapReduce came into existence to handle Big data. Many distributed programming frameworks such as Hadoop, Haloop, SailFish and so on came to process Big data. Cloud computing resources are also utilized by Big data. Stated differently Big data mining and cloud computing work together and complement each other. Big data mining has advantages and many issues or challenges associated with it. Therefore, it is essential to have fair knowledge on Big data, mining Big data, distributed programming frameworks, and cloud computing besides the relationship among them. This paper focuses on achieving this goal. Big data can create value in terms of transparency, experimental analysis, business intelligence, real-time analysis and decision making, and computer-assisted innovative solutions. Big data mining can improve prediction or forecast accuracy that adds big value to businesses [2]. Internet communications, social networks, human digital universe, web search engines are some of the applications that produce Big data. The existing techniques such as machine learning techniques, unstructured data analytics, visualization, data mining, cloud computing, graph and mesh algorithms, and joining algorithms do not scale to processing Big data in Zettabytes [3]. Big data can bring about many advantages such as increasing operational efficiency, strategic planning, better customer service, identifying customer needs, enhanced customer experience, identifying new markets, penetration into new markets, complying with regulations, and others. The knowledge discovery process in big data mining includes phases such as data recording, data cleaning, data analysis, data visualization and interpretation, and decision making. Big data techniques that bring about these advantages can be classified into mathematical techniques, data analysis techniques, and Big data applications. With respect to stream processing Big data tools available are Storm, S4, SQL Stream Server, Splunk, Apache Kafka and SAP Hana [4]. As businesses in the real world are producing huge amount of data and that needs to be harnessed to make strategic decisions, Big data mining plays a vital role in the growth of organizations. Comprehensive business intelligence can be derived by mining Big data. There are many aspects associated with Big data mining. They include MapReduce programming model, distributed programming frameworks like Hadoop, security, privacy preserving data mining, data centres, cloud computing, algorithms and applications for big data analytics. The rise of big data or the arrival of Big data era has forced the world to have infrastructure to mine such data. The infrastructure includes cloud computing resources, distributed programming frameworks and applications that can analyze huge amount of data with various underlying algorithms. With big data mining, it is possible to mine huge data with diverse dimensionality and heterogeneity. Our contribution in this paper is the review of present state of the art of big data, big data mining, algorithms, distributed programming frameworks, cloud and computing and the relationship among them. The remainder of the paper is structured as follows. Section 2 reviews literature on big data and its characteristics. Section 3 throws light into big data mining. Section 4 provides review of distributed programming frameworks. Section 5 focuses on the relationship between cloud computing, big data mining and the underlying programming model and distributed programming frameworks. Section 6 presents algorithms for massive-scale data mining while Sect. 7 concludes the paper besides providing directions for future work.

2 Big Data

Big data refers to huge amount of data with properties such as volume, velocity, and variety. Enterprises in the real world are accumulating data with exponential growth. Such data is termed as big data. It is also characterized y other attributes such as variability, complexity, and value [1]. The knowledge on big data technology and the tools needed to process data besides Big data practices can help individuals to have freedom on the subject. Big data has arrived and its impact will be far-reaching in future when it is coupled with technologies like cloud computing [5].

2.1 Volume

This is an important feature of Big data which indicates that data is very huge and it increases in exponential growth. For instance, social networking applications are producing terabytes of data each day which amounts to such voluminous data which cannot be handled with traditional systems. There is unprecedented growth of data due to globalization ad collaborations among organizations leading to the need for Big data technology.

2.2 Velocity

Velocity attribute of Big data refers to the fact that it deals with data that comes faster from various sources. Stated differently data is accumulated with speed from different sources. For instance, sensor networks continuously sense data and send huge amount of data to base station. The speed at which data is flown is indicated by velocity. Big data technology needs to deal with such data as well.

2.3 Variety

This attribute of Big data deals with data of different kinds. The data that comes from different sources do not belong to a single category. The sources of data include sensor networks, social networks, weblogs, documents, e-mails, news articles, and so on. Handling such diversified data in raw, structured, semistructured and unstructured format is not possible with traditional systems.

2.4 Variability

This property refers to data flow inconsistencies. Data coming from various sources do not necessarily be consistent. There might be sudden increases in the data flow due to certain events that influence social networking websites or other sources causing Big data flows.

2.5 Complexity

This attributes indicates the complexity of Big data since it comes from various sources, contains different data formats, comes with different speed and variability, and needs to be linked or connected and correlated. Thus, Big data is very complex in nature.

2.6 Value

This is the ultimate property of Big data. It refers to the value it adds to enterprises in the real world. Organizations can derive business intelligence (BI) from Big data through a process known as Big data mining. Such BI can help organization to make strategic and well-informed decisions which lead to faster growth of the business. Moreover, Big data mining can provide unbiased and comprehensive BI which increases the power of decision making. Big data is associated with many things. They include data sources, content format, data stores, data staging, and data processing. Figure 2 shows the classification of big data that provides quick glance Big data.

There are many data sources available such as World Wide Web (WWW), machine, sensing, social networking, transactions, and IoT. The Big data is available in different formats such as structured, semistructured, and unstructured. The data stores are in the form of document-oriented, column-oriented, graph based, and key-value. Data staging includes cleaning, normalization and transform. Data processing is of two types namely batch processing and real time process.

3 Big Data Mining

3.1 Need for Big Data Mining

When data is available in abundance surely there are certain latent trends or patterns that can be discovered. Big data mining is the process of discovering comprehensive business intelligence from the bulk of data. This is essential when cross-organization data is involved in the Big data. Since Big data is different from traditional data the traditional approaches are inadequate to handle it. Processing mountains of data for discovering knowledge is always a task that needs specialized infrastructure. Figure 1 illustrates the need for mining Big data.

As shown in Fig. 2, it is evident that blind men having limited view make wrong conclusions. Therefore, considering a part of Big data does not yield accurate BI. Instead, such activity leads to inaccurate knowhow that leads to making wrong decisions. Therefore, it is important to have big data mining in place where the data is Big data. Big data adds big value to businesses provides such data is mined comprehensively for making strategic decisions toward organic and inorganic growth of enterprises in the real world [6].

3.2 Databases to MapReduce

Traditional databases work fine with data that is at rest. They have limitations dealing with Big data which is characterized by volume, velocity, and variety. The open sources databases like MY SQL, Postgress lag behind in providing scalable

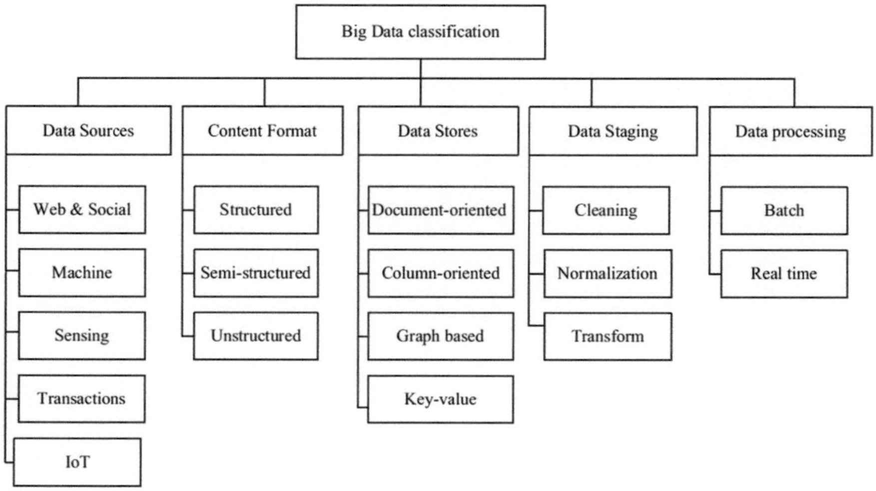

Fig. 1 Classification of big data [17]

Fig. 2 Limited and blind views lead to biased conclusions [32]

solutions when compared with commercial counterparts. They cannot handle massive quantities of data in parallel. Therefore, new programming model such as MapReduce came into existence which can scale well to the Big data that even grows exponentially. Traditional databases have their own advantages as they do consider only relational data. MapReduce or Hadoop are models that support handling Big data with said characteristics [7]. Many tools came into existence that is based on MapReduce. Apache Mahout, Spark, Haloop, Twister, and Daytona are examples that leverage MapReduce programming paradigm to provide better support for Big data mining [7].

3.3 Big Data as a Service (BaaS)

The emergence of cloud computing has led to many services which are available through the Internet in pay per use fashion. In addition to the existing services such as Software as a Service (SaaS), Infrastructure as a Service (IaaS) and Platform as a Service (PaaS), Big Data as a Service (BaaS) is the latest possibility due to the need for big data mining and the computing resources it demands for processing. It encompasses all the existing three service layers making them as Big Data Software as a Service (SaaS), Big Data Infrastructure as a Service (IaaS) and Big Data Platform as a Service (PaaS). The possible analytic techniques that can be included into Big data as a service include user behaviour modelling, recommender systems, insurance, manufacturing process analysis, marketing and sales, QoS prediction, fault tolerance, and performance problem diagnosis [8].

3.4 Issues and Challenges

Big data brings big challenges too. The challenges can be categorized into privacy and security, data access and sharing, storage and processing, analytics, skill requirement, technical challenges such fault tolerance, scalability, quality of data, and heterogeneous data. Privacy and security issues come into the picture in Big data mining as the data can be abused and used for privacy invasion, increased state and corporate control, decreased civil freedoms and invasive marketing. Analytical challenges include dealing with huge data, storage, scope in analysis, and utilization of the results. Since Big data is an emerging technology skill development is an issue [1]. To overcome these issues, Begoli and Horey [9] provide design principles such as supporting a set of analysis methods, support for storing different kinds of data, and making data accessible besides security features. Moreover, domain knowledge is essential for the successful design of such systems [10] besides having the ability to exploit parallel processing power [11], [12]. One of the big challenges of Big data is to mine voluminous data and discover useful information for making accurate business decisions [13].

3.5 Big Data Platforms

Though the research made on Big data is relatively less, there are handful of Big data platforms that can help in processing Big data for real-time processing and batch processing.

Big data platforms support batch processing and real-time processing as presented in Fig. 3. Apache Hadoop is one of the well-known platforms for

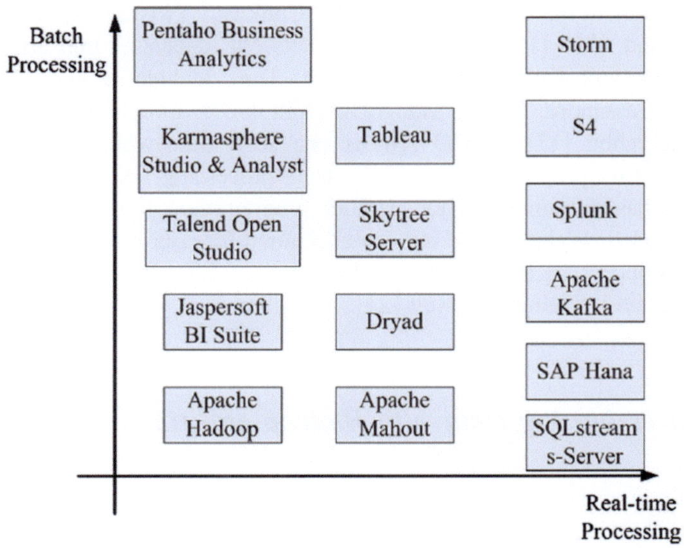

Fig. 3 Platforms for big data processing [4]

MapReduce programming and HDFS. Dryad is another programming paradigm that is based on dataflow graph processing (DGP). Apache Mahout provides collection of machine learning algorithms for intelligent data analysis large-scale applications commercially. Automatic report generation is supported by Jaspersoft BI suite. Another such tool for generating reports is Pentaho business analytics [4]. Skytree Server is a system for data analytics. Tableau provides three tools such as Desktop, Public, and Server for big data processing. Karmasphere is another platform for processing big data faster. Talend Open Studio provides graphical environment for big data processing. S4 is specially designed cloud platform for handling streaming data. Splunk provides real-time business intelligence from machine-generated big data. Apache Kafka is a high-throughput messaging system that makes use of in-memory analytics. SAP Hana is another in-memory analytics platform for scalable batch processing. Google's Dremel is an interactive system for analysis of nested data. Apache drill is another such tool for interactive processing of Big data [4]. Big data mining might have privacy problems when the mining task is outsourced [14, 15]. In [16], there are mechanisms such as K-anonymity, l-diversity to protect published data from being abused. These techniques can avoid attacks on the data from identity disclosure.

4 Distributed Programming Frameworks

Distributed programming frameworks provide the infrastructure that can be used to process big data in the context of cloud computing and distributed file system. Stated differently, these programming frameworks do support MapReduce programming paradigm that can process large volumes of data. The distributed programming frameworks that support MapReduce include Hadoop, Haloop, Hive, Hbase, Madout, Pig, Zookeeper, Spark, Chukwa, Twister, MAPR, YARN, Cassandra, and Avro [17]. There are many NoSQL databases that are used in a cloud environment for big data processing. They include DynamoDB, Redis, Voldemort, Cassandra, Hbase, MangoDB, SimpleDB, CouchDB, BigTable, and Apache Jackrabbit [17]. DisCO [18] is one such framework for collaborative aggregation for practical distributed data processing. Phoenix is another MapReduce-based framework for parallel programming that leverages modern GPU power [19]. Weka tool which is well known for data mining was integrated with MapReduce [20] to support big data mining. Effective frameworks should support data stream mining as well [21].

4.1 Processing Big Data with Hadoop and Haloop

Hadoop is one of the distributed programming frameworks that are widely used in the real world. It is based on the new programming paradigm named "MapReduce."

Bu et al. [22] studied distributed programming frameworks and proposed a new framework known as Haloop which a variant of Hadoop. Haloop extends the capabilities of MapReduce with additional features like task scheduler, loop awareness, and cashing. The frameworks like Hadoop and Haloop provide scalable solutions that help in processing Big data. Hadoop is being used by companies like Google, Yahoo, and Facebook. Different features like caching, indexing, scheduling, and loop aware scheduling of tasks are presented. There are three layers in the architecture. They include application layer, framework layer, and system layer. The file system layer comprises of local and distributed file systems. The local file system participates in indexing and caching mechanisms. The distributed file system is responsible to reflect files located in the distributed environment. The framework layer has important components like task scheduler, loop control, and task tracker. The task scheduler is responsible to schedule tasks in loop aware fashion with the help of loop control while the task tracker keeps track of tasks. Caching and indexing will speed up the process of completing tasks. The application layer has master and slave nodes that are involved in job processing. MapReduce and Haloop are having iterative processing that works differently. The main difference between them is that the loop awareness is missing in Hadoop. Moreover, Haloop also supports caching mechanism that improves the performance of the framework. PageRank algorithm is the common feature of both the frameworks.

4.2 Large-Scale Data Processing with Sailfish

Rao et al. [23] focused on another framework named Sailfish that is based on MapReduce programming paradigm. It can be used to process Big data. Moreover, it has enhanced functionality for MapReduce tasks which are part of MapReduce programming. Its performance is increased by 20% when compared with Hadoop in terms of speed. Its auto-tuning features improve performance significantly. These authors also studied Dryad and Hive and compared with Sailfish. As demonstrated by their work, it is evident that the Map and Reduce tasks are improved in Sailfish [41]. There are flexible features that let the framework to work better in processing Big data. The outputs of Map tasks are given as inputs to Reduce tasks. These tasks work in tandem with its auto-tuning features which enhance the performance of the framework. The performance improvement was proved by using datasets provided Yahoo. As those datasets are benchmark datasets, the evaluation could be made easily. The outputs are aggregated in Sailfish for efficient processing. The framework can leverage the parallel features of the runtime environment. This is again achieved using its famous autotuning process. There are intermediate operations for higher performance. The intermediate steps are also fine-tuned to improve performance. Thus, around 16 TB intermediate data can be handled by Sailfish. Sailfish reduces the computational overhead by supporting batch processing It is evident that the tasks are executed in batches. Using this feature whenever required, the framework can reduce computational overhead. It was made possible by using intermediate processing effectively. I files are used to process data that is processed batch processing.

5 Relationship Between Cloud Computing, Big Data and New Programming Paradigm

Cloud computing has emerged a new computing model that works in distributed environment. Stating differently the cloud computing is an emerging technology that helps users across the world to gain access to a pool of resources through Internet. The services provided by cloud are accessed by users in pay as you use fashion. Virtualization is the technology that made the cloud computing affordable. Cloud resources are unlimited in the sense that they are maintained in such a way that the computing facilities are always scalable. Thus, the cloud computing resources can be used for big data mining in a distributed environment. Cloud data centres and the availability of distributed programming frameworks are the ideal candidates for Big data mining. The new programming paradigm such as MapReduce is the new programming approach for processing Big data. Distributed programming frameworks that support MapReduce or similar kind of programming model are suitable for Big data mining. Therefore, it is possible that the cloud computing, big data, big data mining, distributed programming frameworks like Hadoop with MapReduce go hand in hand. Since cloud computing and Big data are fast growing technologies that complement each other we cannot imagine them growing separately. Big data makes use of cloud resources especially storage. Moreover, Big data analysis can be done by using cloud-based applications which are fast growing [17]. MapReduce programming paradigm that works on top of Hadoop Distributed File System (HDFS), a distributed file system, running in the cloud is capable of processing big data. There is the presence of query engines like Hive and Mahout and so on for processing Big data. The unstructured data is stored and accessed through HDFS. HDFS is the file system that can serve in a distributed environment where thousands of nodes are interconnected [17]. There are many big data cloud platforms such as Google, Microsoft, Amazon, and Cloudera. Google cloud services, Azure, S3 are big data storage facilities of Google, Microsoft, and Amazon, respectively. The big data analytics supported by them are BigQuery, Hadoop on Azure and Elastic MapReduce. Machine learning algorithms associated with those clouds include Prediction API, Hadoop+Mahout, and Hadoop+Oryx. The successful case studies that made use of Big data and cloud include Swiftkey, 343 Industries, redBus, Nokia, and Alacer.

6 Algorithms for Massive-Scale Data Mining

Han et al. [24] proposed two Apriori based algorithms for massive-scale parallel processing of data to obtain frequent item sets and generating association rules. The two algorithms are known as Intelligent Data Distribution (IDD) and Hybrid Distribution (HD). The IDD algorithm minimizes computational overhead besides reducing idle time of processors. The HD combines best features of CD and IDD that scale well to process huge amount of data by aggregating processors and leveraging parallel processing power. The task of finding frequent item sets is done in parallel thus improving speed and cost-effective mining of association rules. Azzini and Ceravolo [25] proposed mechanisms for semantic lifting approaches in building algorithms for Big data mining without compromising requirements in the context of "Big Data". Thus, they paved way for consistent process mining algorithms that keep semantic lifting in mind that takes care of problems such as data types mismatch and values mismatch while using the programming paradigm "MapReduce." Hoi et al. [26] proposed an algorithm for online feature selection (OFS) as part of Big data mining. The algorithms involved in the solution include truncation of OFS and sparse projection. The first one extends perception algorithm using truncation. However, its performance is limited and does not guarantee less number of mistakes. The second algorithm is based on sparsity property. In the second approach, a linear classifier is maintained by online learner which is constantly updated in order to have modified projections. The algorithms are tested in terms of online learning performance and online versus batch comparison. Then the same is evaluated on Big data. The feature selection tasks proved to be effective for online learning applications. Rakthanmanon et al. [27] proposed four novel ideas to work with massive time-series data. They employed an approach known as Dynamic Tim Wrapping (DTW) that can act as a measure in order to work efficiently for big data. The experiments with voluminous data proved that the DTW measure is best used with time series for faster mining of data. Laptev et al. [28] proposed a framework named Early Accurate Result Library (EARL) for processing Big data besides faster estimation of results and accuracy of the same. The API provided by EARL can be used to build new algorithms that can work for Big data analysis. EARL uses Hadoop for error estimation. Zhang et al. [29] proposed an online learning algorithm that mines big data in distributed environment. The learning mechanism contains multiple local learners and an ensemble learner to achieve prediction accuracy. In [30] a review is made on various algorithms that can process huge amount of data. They include R-Tree, R*-Tree, NN Search, Decision Tree Learning, Decision Tree C4.5, GA Tree, and Hierarchical Neural Network. For structural analysis of Big data Kang and Faloutsos, [31] explored many algorithms. They include GIM-V and HEigen. These algorithms are used to unify other graph mining algorithms such as node proximities, PageRank, diameter, Random Walk with Restart and connected components. Big data mining algorithms in distributed environment need to have local learning and model fusion features for different sources of data. Algorithms also should work for mining from incomplete, uncertain and sparse data besides mining dynamic and complex data [32]. Big data mining algorithms need to take care of processing of data from multiple sources besides improving the efficiency of traditional methods [33, 34]. Big data mining algorithms should be able to handle complex relationships in social media data [35, 36]. Chu et al. [37] realized many existing algorithms like neural networks, expectation maximization, Gaussian discriminate analysis, independent variable analysis, Support Vector Machine (SVM), Naive Bayes, logistic regression, k-Means, and linear weighted regression. Information sharing in large scale is one of the goals of information systems. However, there are privacy concerns that are to be addressed by algorithms [38, 39].

7 Conclusion and Future Work

In this paper, we focused on Big data which not only refers to huge amount of data (Volume) but also other properties such as variety, velocity, complexity, variability, and value. Since data is accumulated rapidly with exponential growth, it became Big data. Big data adds big value to business provided harnessing of it for comprehensive business intelligence. Big data provide plethora of advantages. With this, it also brings about issues and challenges. The issues and challenges are related to mining huge amount of data and privacy of data. The traditional data mining algorithms are not sufficient to handle Big data. The traditional programming models also do not work well for Big data mining. To overcome this MapReduce, a new programming paradigm, came into existence. This model can process huge amount of data in distributed environment. It can make use of thousands of servers in order to process a request. It is associated with a distributed file system for storing and retrieval of data. There are many programming frameworks such as MapReduce, Hadoop, Haloop, SailFish, and so on for processing big data in the context of cloud computing. Cloud computing, Big data, Big data mining, distributed programming frameworks have relationships between them. The algorithms meant for Big data mining are to be developed in such a way that they support MapReduce programming with the ability to leverage massive parallel processing power of cloud. This paper has provided present state of the art of Big data, Big data mining, distributed programming frameworks, algorithms and the relationship among them. This research can be extended to enhance existing data mining algorithms to handle big data. Another research direction is to propose and implement new algorithms for massive-scale data using MapReduce programming paradigm in distributed environment.

References

1. Katal, A., Wazid, M., Goudar, R.H.: Big Data: Issues, Challenges, Tools and Good Practices, pp. 213–313. IEEE, Piscataway (2013)
2. Bughin, J., Chui, M., Manyika, J.: Clouds, Big Data, and SmartAssets: Ten Tech-Enabled Business Trends to Watch. McKinSey Quarterly (2010)
3. Kaisler, S.: Big Data: Issues and Challenges Moving Forward, pp. 12–17. IEEE, Piscataway (2013)
4. Philip Chen, C.L., Zhang, C.-Y.: Data-Intensive Applications, Challenges, Techniques and Technologies: A Survey on Big Data. Elsevier. pp. 32–44 (2014)
5. IBM What Is Big Data: Bring Big Data to the Enterprise, http://www-01.ibm.com/software/data/bigdata/, IBM (2012)
6. Jacobs, A.: The Pathologies of Big Data. Comm. ACM **52**(8): 36–44 (2009).
7. Madden, S.: From Databases to Big Data, pp. 32–44. IEEE, Piscataway (2013)
8. Zheng, Z., Zhu, J., Lyu, M.R.: Service-Generated Big Data and Big Data-as-a-Service: An Overview, pp. 12–17. IEEE, Piscataway (2013)
9. Begoli, E., Horey, J.: Design Principles for Effective Knowledge Discovery from Big Data, pp. 12–17. IEEE, Piscataway (2012)
10. Kopanas, I., Avouris, N., Daskalaki, S.: The role of domain knowledge in a large scale data mining project. In: Vlahavas, I.P., Spyropoulos, C.D. (eds.) Proceedings of the Second Hellenic Conference AI: Methods and Applications of Artificial Intelligence, pp. 288–299 (2002)
11. Luo, D., Ding, C., Huang, H.: Parallelization with multiplicative algorithms for big data mining. In: Proceedings of the IEEE 12th Int'l Conference Data Mining, pp. 489–498 (2012)
12. Shafer, J., Agrawal, R., Mehta, M.: SPRINT: a scalable parallel classifier for data mining. In: Proceedings of the 22nd VLDB Conference (1996)
13. Rajaraman, A., Ullman, J.: Mining of Massive Data Sets. Cambridge University Press, Cambridge (2011)
14. Lorch, J., Parno, B., Mickens, J., Raykova, M., Schiffman, J.: Shoroud: ensuring private access to large-scale data in the data center. In: Proceedings of the 11th USENIX Conference File and Storage Technologies (FAST '13) (2013)
15. Schadt, E.: The Changing Privacy Landscape in the Era of Big Data. Mol. Syst. **8**, article 612 (2012)
16. Machanavajjhala, A., Reiter, J.P.: Big privacy: protecting confidentiality in big data. ACM Crossroads **19**(1), 20–23 (2012)
17. Hashem, I.A.T., Yaqoob, I., Anuar, N.B., Mokhtar, S., Gani, A., Khan, S.U.: The Rise of "Big Data" on Cloud Computing: Review and Open Research Issues, vol. 47, no. 1, pp. 98–115. Elsevier, Amsterdam (2015)
18. Papadimitriou, S., Sun, J.: Disco: distributed co-clustering with map-reduce: a case study towards petabyte-scale end-to-end mining. In: Proceedings of the IEEE Eighth Int'l Conference Data Mining (ICDM '08), pp. 512–521 (2008)
19. Ranger, C., Raghuraman, R., Penmetsa, A., Bradski, G., Kozyrakis, C.: Evaluating MapReduce for multi-core and multiprocessor systems. In: Proceedings of the IEEE 13th Int'l Symposium High Performance Computer Architecture (HPCA '07), pp. 13–24 (2007)
20. Wegener, D., Mock, M., Adranale, D., Wrobel, S.: Toolkit-based high-performance data mining of large data on MapReduce clusters. In: Proceedings of the Int'l Conference Data Mining Workshops (ICDMW '09), pp. 296–301 (2009)
21. Zhu, X., Zhang, P., Lin, X., Shi, Y.: Active learning from stream data using optimal weight classifier ensemble. IEEE Trans. Syst. Man Cybern. Part B **40**(6), 1607–1621 (2010)
22. Bu, Y., Howe, B., Balazinska, M., Ernst, M.D.: HaLoop: Efficient Iterative Data Processing on Large Clusters, pp. 1–12. IEEE, Piscataway (2010)
23. Rao, S., Ramakrishnan, R., Silberstein, A.: Sailfish: A Framework For Large Scale Data Processing, pp. 1–14. Microsoft, USA (2012)

24. Eui-Hong (Sam) Han, George Karypis, Member, IEEE, and Vipin Kumar, Fellow, IEEE: Scalable Parallel Data Mining for Association Rules, vol. 12, no. 3, pp. 25–34. IEEE, Piscataway (2000)
25. Azzini, A., Ceravolo, P.: Consistent Process Mining Over Big Data Triple Stores, pp. 25–34. IEEE, Piscataway (2013)
26. Hoi, S.C.H., Wang, J., Zhao, P., Jin, R.: Online Feature Selection For Mining Big Data, pp. 12–17. ACM, New York (2012)
27. Rakthanmano, T.: Addressing Big Data Time Series: Mining Trillions of Time Series Subsequences Under Dynamic Time Warping, pp. 23–33. ACM, New York (2013)
28. Laptev, N., Zeng, K., Zaniolo, C.: Very Fast Estimation for Result and Accuracy of Big Data Analytics: The EARL System, pp. 23–33. Springer, Heidelberg (2013)
29. Zhang, Y.: A Fast Online Learning Algorithm for Distributed Mining of BigData, pp. 213–313. IEEE, Piscataway (2012)
30. Yadav, C., Wang, S., Kumar, M.: Algorithm and approaches to handle data—a survey. International Journal of Computer Science and Network **2**(3), 12–17 (2013)
31. Kang, U., Faloutsos, C.: Big graph mining: algorithms and discoveries. IEEE **14**(2), 25–34 (1998)
32. Wu, X., Zhu, X., Wu, G.-Q.: Data mining with big data. IEEE **26**(1), 97–107 (2014)
33. Chang, E.Y., Bai, H., Zhu, K.: Parallel algorithms for mining large-scale rich-media data. In: Proceedings of the 17th ACM Int'l Conference Multimedia (MM '09,) pp. 917–918 (2009)
34. Wu, X., Zhang, S.: Synthesizing high-frequency rules from different data sources. IEEE Trans. Knowl. Data Eng. **15**(2), 353–367 (2003)
35. Chen, Y.-C., Peng, W.-C., Lee, S.-Y.: Efficient algorithms for influence maximization in social networks. Knowl. Inf. Syst. **33**(3), 577–601 (2012)
36. Zhao, J., Wu, J., Feng, X., Xiong, H., Xu, K.: Information propagation in online social networks: a tie-strength perspective. Knowl. Inf. Syst. **32**(3), 589–608 (2012)
37. Chu, C.T., Kim, S.K., Lin, Y.A., Yu, Y., Bradski, G.R., Ng, A.Y., Olukotun, K.: Map-reduce for machine learning on multicore. In: Proceedings of the 20th Annual Conference Neural Information Processing Systems (NIPS '06), pp. 281–288 (2006)
38. Howe, D., et al.: Big data: the future of biocuration. Nature **455**, 47–50 (2008)
39. Huberman, B.: Sociology of science: big data deserve a bigger audience. Nature **482**, 308 (2012)
40. Wu, X., Zhu, X., Wu, G.-Q., Ding, W.: Data Mining with Big Data, pp. 23–33. IEEE, Piscataway (2013)
41. Mervis, J.: U.S. science policy: agencies rally to tackle big data. Science **336**(6077), 22 (2012)

Content-Based Social Network Aggregation

Charu Virmani, Anuradha Pillai and Dimple Juneja

Abstract Collecting and collating the data from different Social Network sites (SNS) at one place is a concern of today's scenario. A social network aggregator performs this task by drawing information together from multiple SNS's like Facebook, Twitter, and LinkedIn. It also unifies multiple social networking profiles into a single profile. The content presented by these aggregators appears in real time and there is easy switching between the multiple networks. There are many existing Social network aggregators based on the content and features. However, this paper focuses on the content-based architecture of social network aggregator that provides integrated search across multiple social networks.

Keywords Social network · Content aggregator · Social network aggregator

1 Introduction

Social Web is a matrix of interconnected relations and connection that connect people. Social network aggregator is tasked to aggregate social data across multiple social networking sites. A typical social network aggregator does not require separate logins in individual social networks once each has been a trusted site in the aggregator. It is an excellent way to retrieve required information from a different

C. Virmani (✉) · A. Pillai
YMCAUST, Faridabad, India
e-mail: charu.fet@mriu.edu.in

A. Pillai
e-mail: anuangra@yahoo.com

D. Juneja
DIMT, Faridabad, India
e-mail: dimplejunejagupta@gmal.com

A.K. Saini et al. (eds.), *ICT Based Innovations*, Advances in Intelligent Systems and Computing 653, https://doi.org/10.1007/978-981-10-6602-3_18

social network of the user. The aggregation of social networking sites is achieved by switching on an activity stream [1] which user has an option to start and stop. An activity stream is a snapshot of latest activities performed by an individual in a particular social network. For Example, tweets of Twitter.

APIs provided by the different social networks are used to aggregate the social data. An API can access an event initiated by the user from any platform, by authorizing the social-aggregation platform, normally by specifying user-id and password of the social network site to be integrated. This is similar to the open id (a single sign-on system) [2] concept. From the viewpoint of SNS, the 'walled garden' problem is gaining access to user's data. This means that there is no open platform to access the social data across networks.

With lot many SNSs growing in the present scenario, there is a high probability that one user may have accounts with multiple SNSs. According to Anderson Analysis [3], at least 1 over 11,000 SNS users, there is an overlap of users in SNS's. The problem will worse if a user has multiple profiles in same SNS.

Another issue in aggregating social data is to visualize the content of the data [4] which may be social bound or relationship, common interest, location, explicit declaring friends, followers, etc. So, there is a need of understanding various key features and configurable option in popular SSN, and try to make them configurable so that a personalized aggregated solution is available to the user. Collecting social data from multiple social networking sites and providing analysis of relationships and data is the primary concern of content aggregators. It analyzes feature set and correlates user's hobbies, interests, etc. using user abstraction modeling on the basis of the previous preference/selection of data by the user [5, 6].

In this paper, a content-based social network aggregator has been proposed which will attempt to resolve the 'walled garden' problem. It also has an ability to integrate various social networks like Facebook and Twitter. The aim is to effectively aggregate the social data in activity streams. These activity streams are based on the user content as generated on each SNS and group it as activity stream.

The paper explains the content-based SNA as following; Section 2 provides a summary of existing aggregators, Section 3 provides the insight to the proposed solution, Section 4 provides the implementation and demonstration of the prototype developed; Section 5 provides conclusion and further scope of the research.

2 Literature Review

Content Aggregators, Comparison Analytics, Relationship Aggregation, and Process Aggregation [6] are typical ways to integrate social data across multiple social networking sites. However, a lot of research has been done on aggregators to provide integration of social data. Some of the important once are

2.1 Hoot Suite

Hoot suite [7] is a software platform for providing an experience of web dashboard to the business companies to improve upon the marketing promotions, identifying the new users, and their needs to dispense target messages by applying various social networking strategies. It enables users to manage, schedule post, bookmark, handhelds integration, RSS feed, and publish the post to multiple social media accounts at one place. With Hootsuite, users can post updates, connect with the client base, and review responses on more than 35 popular social networks, such as: Foursquare, Facebook (including Events, Groups, Profiles, and Fan Pages), Twitter, LinkedIn (including Pages, Profiles, and Groups), etc.

2.2 People Aggregator

People Aggregator [7] is a service-based social network aggregator. It amalgamates distributed profiles of users spread over several social networks and provides a central service to manage all content like blogs, media galleries, forums, etc. It also disambiguates the identity of the user using ‘connective tissue’ between the profiles to provide a marriage of different profiles into one unique profile. It also provides a unique summary of the user credentials to the recruiter which is more trustworthy source of information than the conventional networks. People Aggregator access photos via API’s provided and establish a connection between two systems.

2.3 SocConnect

SocConnect [8] uses personalized suggestions and semantic contexts for aggregating the social data across multiple social networking sites. It models an intelligent system to provide a personalized suggestion of friend’s activities using machine learning techniques. It learns the preference of the user and suggests new unread information to the user on the basis of their historic preference. Users can emerge a new group of friends by combining friends from various networks and can also rate the activities of their friends.

2.4 Flock

Flock [7] provide a personal experience of web by integrating the status updates and other social data like photos, friend’s update, etc. from other popular social networking sites like Facebook, MySpace, Twitter, etc. It run polls for the feedback and decision-making, provides sharing of rich notes and automatic updates from 40+ tools, sends reminders and thus helps in taking decision faster.

A social network aggregator aggregates the social network members and social data to share social network activities. The very rationale for have an aggregation is to let the user have a one unified window to manage his social interaction and activities without hopping on each SNS separately. All content appears in real time (or abstracted to be appearing), which eliminates the need to hop from one social SNS to other. Justification of having aggregation lies in the fact that not every SNS can be the best place for a user having varies interest and hobby areas. Social Data across multiple sites can hence be integrated on a common platform or protocols [9] as a representation of abreaction of user's preferences. Since each SNSs have their own syntaxes and formats for representing social data. Open web communities have developed standard representations of social data; of these most important methods is OpenID, Activity stream.

These standards provide the basic foundation and building blocks of social aggregators. IT companies and SNS have implemented these standards. For example, Facebook's News Feed is an activity stream.

Keeping context of the data in integrating social networks is another point of consideration. Normally in most popular social networks, the contexts used are social bound, relationships, common interest, etc. [1].

3 Proposed Solution

The main idea behind designing a content-based social network aggregator (CBSNA) is to aggregate multiple social networking sites and fetch social activities (content and features) at one place. In this section, we will describe the schema proposed for social network aggregator.

3.1 Data Flow and Aggregation Representation

The data flow for the aggregator is shown in Fig. 1. It explains the integration of social activities across networks and the role of the proposed aggregator in the same.

Each SNS would have set of social activities which can be aggregated and collated at one place. Typical shared activities are blogs, newsfeed, applications, notifications, contacts, etc. The proposed Aggregator would manage single signing in for all SNS connected by an open OAuth protocol as discussed in previous section.

3.2 Basic Schema of CBSNA

Figure 2 represents the basic schema for the proposed aggregator. It contains five entity types namely social activity, SNS aggregator, SNS member, Tags, and groups. The SNS aggregator entity is the heart of the schema that represents the

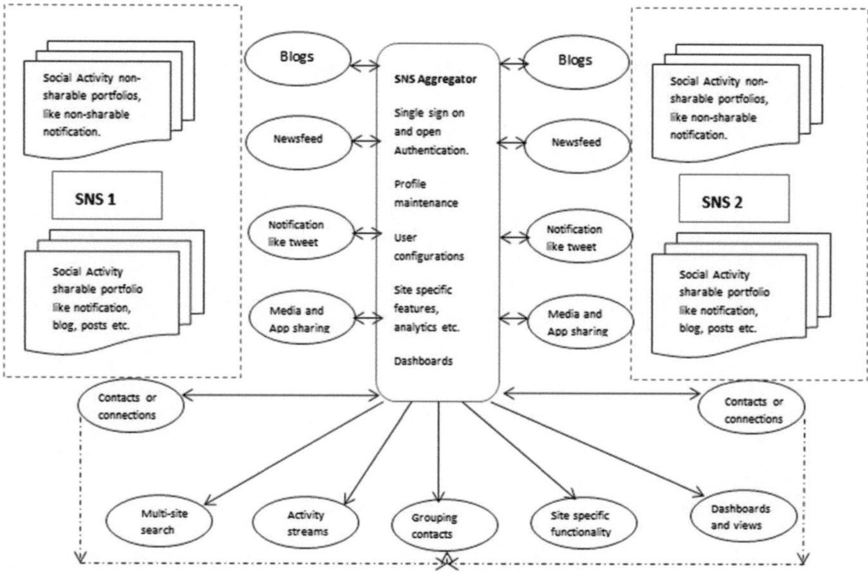

Fig. 1 Data flow diagram of CBSNA

data or the input is sourced by various social activities initiated by user profile in a Social network site. The aggregator will connect with the activities of the SNS member or the group to which the user belongs in that SNS. CBSNA profile will contain user-specific interest and data and the available functionalities. There is another important attribute called source that links the aggregator to the site from which the social activity is generated. The data updating will depend on the refresh rate of the aggregator. The Social Activity entity is defined as the actual social activity which is generated by the user and which is in the shared portfolio of the SNS in concern. Tag represents a user-generated mark. These represent contextual information of social data.

3.3 *Functionalities*

The Aggregator consists of various features:

1. Profile management to maintain the single sign on for different social networking sites. The user needs a onetime authentication for each SNS; thereafter the data from those sites becomes trusted. User needs to login to the aggregator and the data from these trusted sites will bypass the authentication.
2. User configurations and settings like preferences and interests can be configured in the proposed SNA so that user can subscribe to a required set of activities, he wish to see as an activity stream. Various activity streams can be tweets, blogs, news publications, etc.

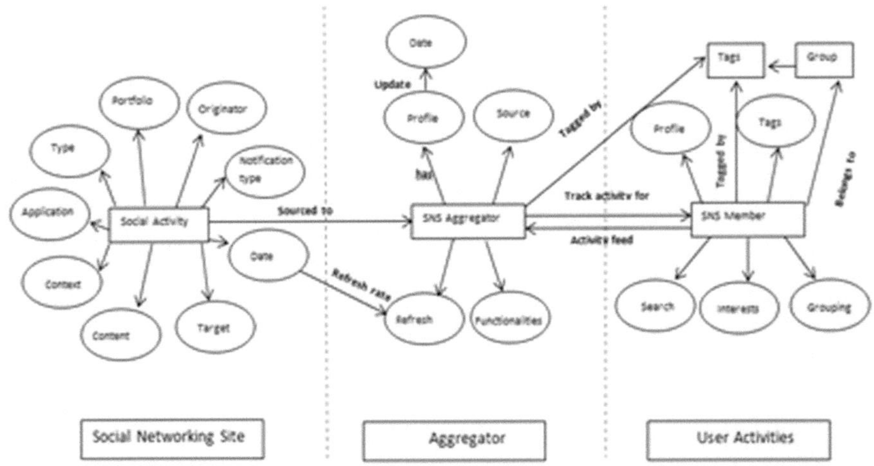

Fig. 2 Schema design of CBSNA

3. Dashboards to display the data based on the site of integrated for all to have a unified view of all the configured and available functionalities for each SNS.
4. Updating comments/feeds by regularly refreshing the feeds from SNS in order to have a real-time access to the information.
5. Grouping contacts will groups various contacts from different sites at one place. The profile with same email id which has a presence in multiple sites will be encapsulated together to represent one contact. However to distinguish the contact notification of the same across networks a tag field is used that marks the source of notification or social data.
6. Another unique feature of the proposed solution is a multi-site search functionality which essentially searches a keyword (contact or information) across many configured SNS for that user. This means if one can search for the data from various sites for which the search is configured. For example, User can search other users of multiple networks at one place, i.e., Ram can be searched in Facebook, Twitter, and LinkedIn as well. A similar search can be made for groups as well.

4 Implementation and Demonstration

In this section, content-based social network aggregator is implemented and various functionalities. The implementation platform is VB.NET implemented on.NET framework 4.6. Visual Basic version used for the development is VB 10. IDE used is Visual Studio 2013. The Functionalities is as below.

4.1 Single Sign on and Authentication

OAuth protocol [10] is used to authenticate each SSN configured on the Social network aggregator to track social activities being initiated from these sites. OAuth [10] which is essentially an open authorization protocol is supported by most of the SNS's [11]. SNS's can be added as trusted partners [12] when user login to the aggregator thereby providing a 'single sign on' ability. For the purpose of demonstration Facebook, LinkedIn, and Twitter is used in this prototype.

After the user has signed in the aggregator he/she can see the social networks in his/her machine. Three network tabs have been provided as default. Authentication to various SNS can be provided using the single sign on functionality. This is achieved by trusting the SNS.

For each SNS (Facebook, Twitter, and LinkedIn) in this case, the user has to click on the tab to authorize them for this SNS. This essentially means that the aggregator is actually marking the said site as 'trusted' and will not ask for the password again to login.

4.2 Content-Based Data Presentation and Social Activity Tracker

The aggregator will have a list of features pertaining to each SNS which appears for that particular site tab. User can customize the content that he wish to share and publish for that site in the aggregator. As discussed these are based on the activity stream and attempt to capture the stream as created by the user, as soon as user click on the feature, the streams gets closed and next stream open up.

A user can select the feature to display the relevant feed or information for Facebook as demonstrated in Fig. 3. The respective SNS to check the feeds/comments can be selected from the tabs.

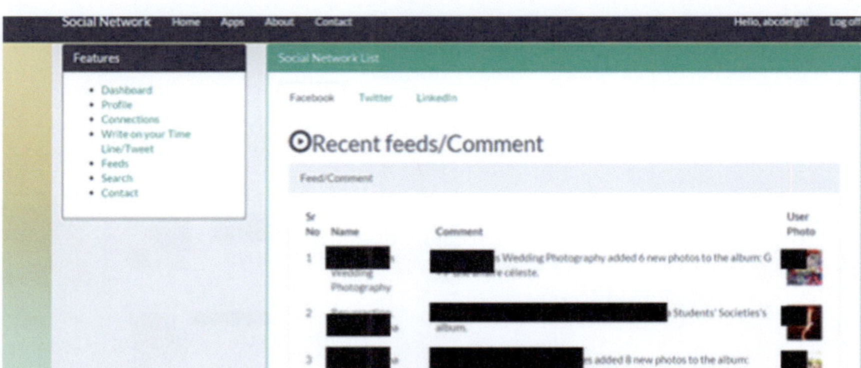

Fig. 3 Activity stream of feeds/comments

4.3 Real-Time Update by Writing Post on Users' Time Line

The Aggregator provides the real-time update of user's status or notification input to the main SNS as well. This has been handled as activity stream. An update in form of comment can be posed simultaneously in multiple sites by selecting the posed on checkboxes.

4.4 Integrated Contacts

This is a unique feature that is not present in most of the Aggregators: integrated contacts. The fact that contact needs to be put together in one place is encapsulated and abstracted view of the SNS integrator as a whole. The system will also sense those profiles which have an overlap in more than one SNS and put them as one entity/contact as shown in Fig. 4.

4.5 Integrated Search

This is another unique feature that has been provided in the prototype, i.e., multi-site graph search. A contact, keyword (place, contact for a demonstration in this prototype) can be searched from multiple sites (configurable) and provides the integrated results. This is essentially query-based graph search as shown in Fig. 5.

The proposed solution enables an end-to-end social networking experience by integrating social data across multiple sites in one dashboard. With a user-friendly interface, an attempt has been made to provide an abstraction of social activity by

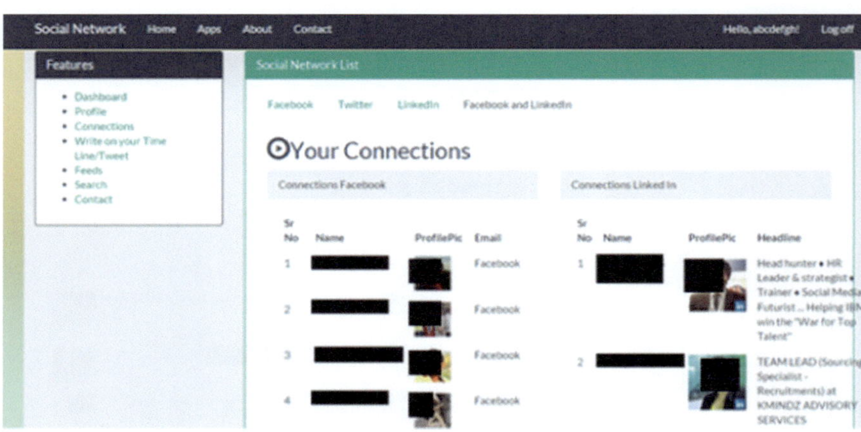

Fig. 4 Integrated contacts

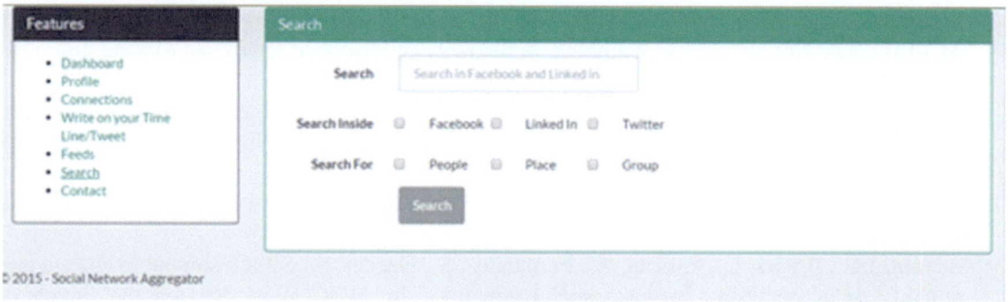

Fig. 5 Integrated search

grouping the contacts, multi-site search to an extent. The proposed system has opened doors for the 'walled garden approach' thereby extracting information across multiple SSN's. This is a step ahead to integrate the social information.

5 Conclusion and Future Scope

Social network aggregators collates information and tracks the social activities from various social networks. The proposed prototype demonstrated provides a comprehensive solution of SNS integration with some unique user-friendly and powerful features like integrated profile management and integrated search capabilities.

To summarize, an aggregator is developed that integrates several social web sites together and extracting the useful information among multiple social networks. This has given an edge over the typical activity stream bases social activity implementations and is a step ahead to integrate the social data. Proposed aggregator has abstracted few features as an integrated solution with contacts and search capabilities, however, further research can be extended on abstracting other features like an abstracted friend, groups, communities, blogs, etc.

References

1. Facebook Users Revolt, Facebook Replies. TechCrunch. Retrieved 11 July 2010
2. Jennings, D.: Aggregation of Data Across Social Networks, alchemi.co.uk. Retrieved (2008-04-10)
3. Brandtzaeg, P., Heim, J.: Why people use social networking sites. In: Ozok A.A., Zaphiris P. (eds.) Online Communities and Social Computing, Chap. 16, vol. 5621, pp. 143–152. Springer, Heidelberg (2009)
4. Ereteo, G., Buufa, M., Gandon, F., Leitzelman, M., Limpens, F.: Leveraging social data with semantics. In: Proceedings of the W3C Workshop on the Future of Social Networking (2009)
5. Catanese, S., De Meo, P., Ferrara, E., Fiumara, G.: Analyzing the Facebook friendship graph. In: Proceedings of the 1st Workshop on Mining the Future Internet, pp. 14–19 (2010)

6. Leskovec, J., Faloutsos, C.: Sampling from large graphs. In: Proceedings of the 12th ACM IGKDD International Conference on Knowledge Discovery and Data Mining, pp. 631–636 (2006)
7. Virmani, C., Pillai, A., Juneja, D.: Study and analysis of social network aggregator. In: International Conference on Optimization, Reliability, and Information Technology (ICROIT), pp. 145–148 (6–8 Feb 2014)
8. Zhang, J., Wang, Y., Vassileva, J.: SocConnect: a personalized social network aggregator and recommender. Inf. Process. Manage. **49**(3), 721–737 (2013)
9. Breslin, J.G., Bojārs, U., Passant, A., Fernández, S., Decker, S.: SIOC: content exchange and semantic interoperability between social networks. In: W3C Workshop on the Future of Social Networking. Barcelona, Spain, 15–16 Jan 2009
10. Adomavicius, G., Tuzhilin, A.: Toward the next generation of recommender systems: a survey of the state-of-the-art and possible extensions. IEEE Trans. Knowl. Data Eng. **17**(6), 734–749 (2005)
11. Web Authorization Protocol (OAuth). IETF. Retrieved 8 May 2013
12. Chen, J., Nairn, R., Nelson, L., Bernstein, M., Chi, E.: Short and tweet: experiments on recommending content from information streams. In: CHI '10: Proceedings of the 28th International Conference on Human Factors in Computing Systems, pp. 1185–1194 (2010)
13. Heckmann, D., Schwarzkopf, E., Mori, J., Dengler, D., Kroner, A.: The user model and context ontology gumo revisited for future web 2.0 extensions. In Proceedings of the International Workshop on Contexts and Ontologies: Representation and Reasoning (2007)

Collaborative Filtering-Based Recommender System

Sangeeta and Neelam Duhan

Abstract Recommender systems have changed the way people find products, information, and services on the web. These kinds of systems study patterns of behavior to know someone's interest will in a collection of things he has never experienced. Collaborative filtering is a popular recommendation algorithm that works to find user's interest patterns and recommendations based on the ratings or behavior of other users or target user in the system. The assumption behind this method is to find a user with similar interest to the active user and use his/her preference for recommendation to the active user. But several issues exist in the kind of method. For example, accuracy, sparsity, and cold start. In this paper, an improved recommendation technique is proposed to address the issues identified.

Keywords Collaborative filtering · Recommender systems · Similarity computation · Item groups

1 Introduction

World Wide Web is tremendously increasing day by day. To find some particular information, a user need to fire a query about what he/she wants to know and then search engine gives suitable results. But what if a user actually doesn't know what information he/she wants to get or a user is not able to fire a reasonable query. This problem is solved by recommender systems. Recommender systems find user interest by analyzing similar users in the system. There are various techniques present in the literature for implementing recommender system [1–3]. Most famous techniques are collaborative filtering and content-based filtering. Collaborative fil-

Sangeeta (✉) · N. Duhan
Department of Computer Engineering, YMCA University of Science & Technology
Faridabad, Faridabad, India
e-mail: sangeeta2yadav@gmail.com

N. Duhan
e-mail: neelam_duhan@gmail.com

A.K. Saini et al. (eds.), *ICT Based Innovations*, Advances in Intelligent
Systems and Computing 653, https://doi.org/10.1007/978-981-10-6602-3_19

tering [4, 5] uses ratings of products provided by the user for prediction. Content-based [6] recommender systems analyze item feature to find the predictions for an individual user. These techniques suffer from a lot of problems, some of which are sparsity, accuracy, cold start problem. These problems are solved by the proposed technique given in next section.

1.1 Collaborative Filtering

It is most familiar and widely used method for the recommendation. Collaborative techniques [7, 8] well work for complex objects, such as movies, music, online products, etc. These techniques make use of user-item rating matrix to compute predictions. These systems aggregates rating on items by a no of users and finds the similarity based on these ratings [9]. For example, Javawock is a recommender system that helps to provides useful Java components (library class files) to a developer based on the collaborative filtering. If a programmer gives an unfinished Java program to javawock, it tries to find Java library class files used in the provided program. Then, the system recommends to the developer Java library class files that were used in the similar programs but were not used in the developer's program [10]. Feng Xie and his team propose several methods to learn more accurate item similarities by minimizing the squared prediction error. This optimization problem is solved using Stochastic Gradient Descent [11].

1.2 Content-Based Filtering

Content-based algorithms make use of item or user profiles and features for recommendation [12]. These systems find the related items based on the items features. Content-based algorithms analyze items description or user profile to find the items of individual's interest. For example, Content REcommendation System is based on the content-based filtering technique. "This system collects and mines the private data of the user at the client side, discovers, stores and updates private Dynamic User Profile (DUP) at the client side [13].

1.3 Hybrid Recommendation Methods

Hybrid recommenders [7, 14] use a combination of collaborative and content-based filtering methods. This removes shortcomings of both methods and gives best results. Wencai Du proposed recommendation algorithm to improve the accuracy of similarity and recommendation quality, taking attribute theory as a theoretical basis [15].

2 Proposed Recommender System

The proposed system is based on the collaborative filtering technique that is used to provide item recommendations to a target user by analyzing its interest. The detailed architecture of the purposed system and explanation of its functional components is given in the subsequent sections.

2.1 The Proposed System Architecture

The detailed system architecture is outlined in Fig. 1 along with all functional components. In the proposed system, User interacts with an online interface by signing in or logging in. All the interaction of following components is given in subsequent section.

The system handles the cold start problem both for the insertion of the new user and for a new item in the system, i.e., it handles how recommendation is done to new user and how new item is recommended to users which are explained in detail in later sections.

2.2 Data Structure Used for Proposed System

A large data needs to be processed for the proposed system for which some data structures are being employed. Here are the details of data structures used in the proposed system.

Movie repository. It contains user-movie matrix A of order $[M \times N]$ and Movie-genre matrix B of order $[N \times K]$ where M is the number of users who visited the site, N is a number of movies in the database and K represents the genres like comedy, action, and romance.

Group rating matrix. Movie repository stores group rating matrix generated by group rating generator. Group rating matrix G of order $[M \times K]$, where M is

Fig. 1 Architecture of proposed system

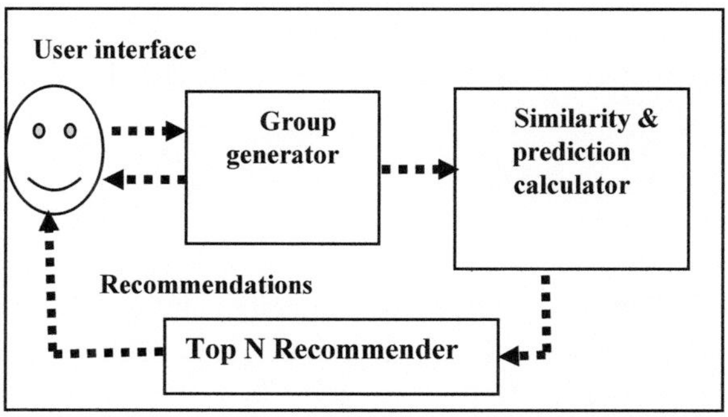

number of users and K is a number of groups generated by group generator. $[G_{ij}]$ represents a rating of jth group given by ith the user. Values of the matrix are the computed using the formula given in (1).

$$R_{u,C} = \frac{\sum_{i \in 1}^{C} R_{u,i}}{\left|R_{u,i}\right|} \quad (1)$$

where $R_{u,C}$ is the rating of Cth cluster for user u. i represents movies in Cth cluster. $R_{u,i}$ denotes rating of ith movie given by user u. $\left|R_{u,i}\right|$ denotes the number of movies in Cth cluster rated by user u. The value of $R_{u,C}$ ranges in [1, 5].

2.3 Functional Components of Proposed System

The description of functional components of the proposed recommender system is given below.

(a) **Group generator**

Group generator generates groups of movies with similar genres. It uses movie-genre matrix to generate different groups. For example, all movies with comedy genre will be clubbed in a group. There are 17 Groups based on the genre are present in actual system implementation. One movie can be presented in many groups based on their number of genre present in that movie. Group generator produces group rating matrix by reading user-movie matrix A and groups of movies and form a matrix which gives rating given by the particular user to a group. If user rating of a particular group is higher as compared to other groups than the user's interest can be identified with in those groups. Group matrix is produced by averaging the ratings of similar movies given by a user which is given by (3).

(b) **Similarity and prediction calculator**

The amount of correlation between the users needed to calculate for similarity computation. Correlation between any two users is computed using Pearson-correlation formula [16] which gives the efficient result. In statistics, the value of the correlation coefficient varies between +1 and −1. When the value of the correlation coefficient lies around ±1, then it is said to be a perfect degree of association between the two variables. The similarity is computed as given in (2).

$$\mathrm{Sim}(u,v) = \frac{\sum_{i \in I} \left(R_{u,i} - \overline{R_u}\right)\left(R_{v,i} - \overline{R_v}\right)}{\sqrt{\sum_{i \in I} \left(R_{u,i} - \overline{R_u}\right)^2} \sqrt{\sum_{i \in I} \left(R_{v,i} - \overline{R_v}\right)^2}}, \quad (2)$$

where Sim(u, v) represents similarity between user u and v. $R_{u,i}$ denote the rating of ith group by user u. $R_{v,i}$ denotes the rating of ith group given by user v. $\overline{R_u}$ and $\overline{R_v}$ are average rating of ith group given by user u and v, respectively. I is set of all users.

Predictor calculates weights of a movie for the recommendation. It uses traditional formula to compute prediction. Predicted values range from 1 to 5. The prediction computation formula [17] is given as (3).

$$P(u,i) = \overline{R_u} + \frac{\sum_{v \in NI} \text{sim}(u,v) * R_{v,i}}{\sum_{v \in NI} |\text{sim}(u,v)|}, \tag{3}$$

where sim(u, v) represents similarity between user u and v calculated from (5). $R_{v,i}$ represents rating of ith item given by user v, P(u, i) gives prediction of item i for target user u. NI is the set of movies which has not been rated by the target user. Predicted values are stored in array list in descending order. Now top 10 values is selected for recommendation to the target user. Value of N can be altered according to requirement. These top N values are then linked to movie repository and corresponding movie information is retrieved and displays to the target user through the online interface.

3 Snapshot of Implemented System

There are some screen shots given which will help the user to show the results.

Figure 2 shows the online interface of a movie recommender system. By authorizing the user (which is not shown here because of simplicity) it can give its

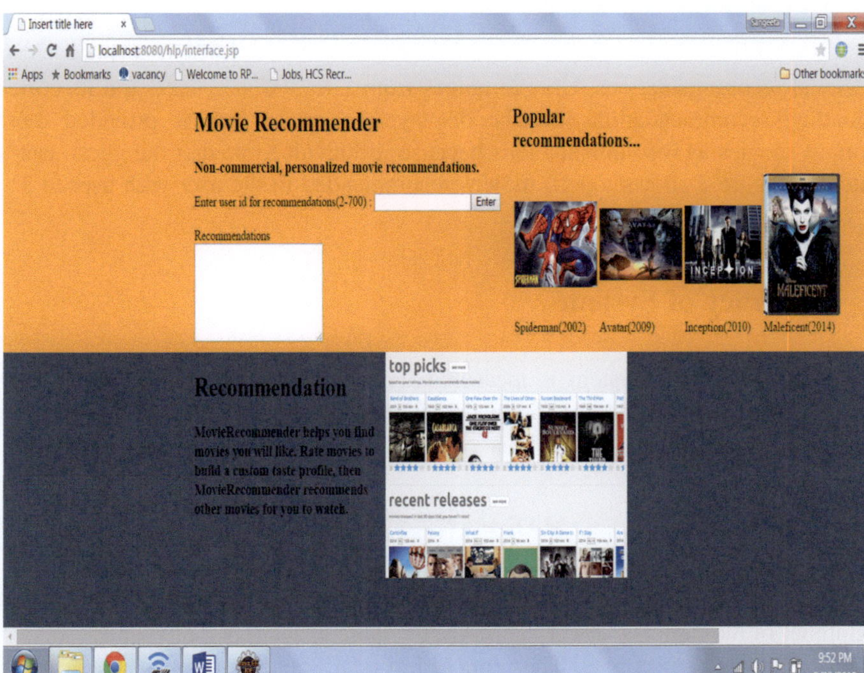

Fig. 2 Interface of online recommender system

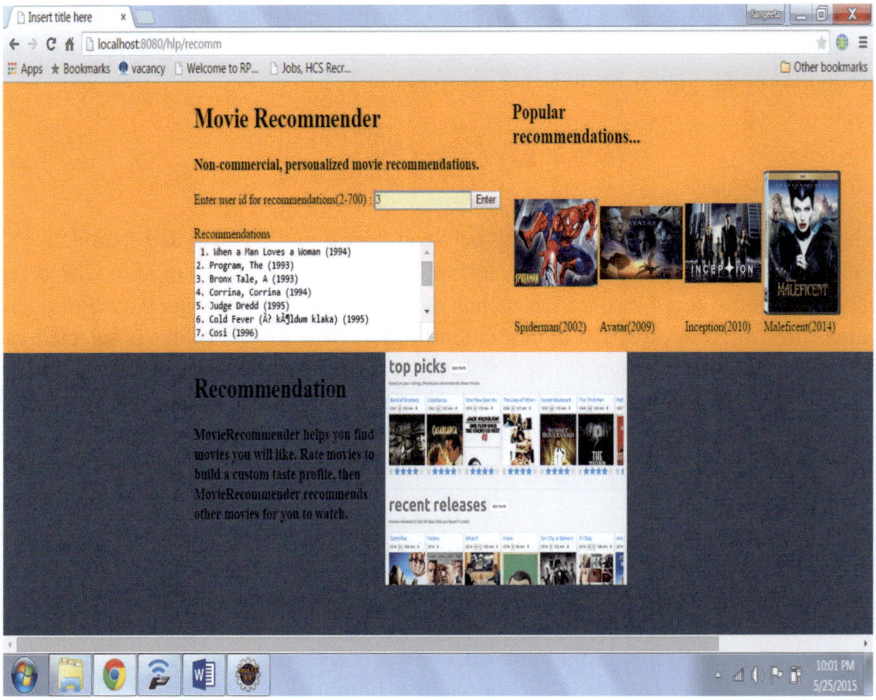

Fig. 3 List of movies for an individual user

personal recommendations. User id ranges from 1 to 700 is entered and its corresponding recommendations will be displayed. A guest user is provided most famous movies as recommendations because it difficult to predict the guest user's interest. Figure 3 shows top 10 movies recommended to the user with user id 3.

4 Accuracy of System

The accuracy of the system is measured with the help of mean absolute error formula [4]. MAE is a measure of the deviation of recommendations from their true user-specified values. Formally, it is given as (4).

$$\text{MAE} = \frac{\sum_{i=1}^{N} |p_i - q_i|}{N}, \tag{4}$$

where p_i is the predicted rating of ith movie, q_i is actual rating of ith movie and N is a number of movies. The lower the MAE, the more accurately the recommendation engine predicts user ratings. The proposed system is checked for accuracy using this measure which shows that that the system is 97% accurate in terms of its predicted rating for the user.

5 Issues Solved by Proposed System

New item problem. When a new item is inserted in the system, it does not have any prior rating by the user so, it cannot be recommended to the users. This is called new item problem.

Solution: When a new movie is added to the system, it is checked with its type of genres and added to the groups of movies with similar genres. So, when a particular group of the movie is recommended to the user, the new item is also recommended with them.

New user problem. When a new user is inserted in the system, the user has not rated enough movies so that his/her interest can be calculated. So it is difficult to recommend such a user. This is known as new user problem.

Solution: A new user is added to the system in user-movie genre matrix A. For a recommendation to such a user, top-n predictor calculates most famous movies rated by the user by computing average rating of a movie. These predictions are recommended to the new user until they rate enough movies to find their interest and personal recommendation is given after that.

Accuracy. Quality of recommended items is an important issue in recommender systems. Recommender system should provide maximum accuracy.

Solution: Proposed system is measured with the metric MAE to check the accuracy of the system. Experimental results show that system is more accurate than other techniques.

Sparsity. Collaborative filtering considers user-item matrix A for user/item similarity computation. This matrix is always sparse. Values rated by users are very less as compared to number of movies. So, it effects the quality of recommendation.

Solution: Proposed system used Group Rating matrix for similarity computation which is not sparse and give efficient results.

6 Conclusion and Future Work

The proposed system works on movies domain. The proposed recommender system forms movie groups based on the genre information which is downloaded from movielens website. These groups solve new item and sparsity problem. Recommendations are made by computing user to user similarity with the help of Pearson correlation method and then predictions are computed. Top 10 recommendations are displayed to the user. This system solves various issues like sparsity, accuracy, cold start, etc.

In Future Groups of movies can be improved using various features of movie, such as director, release date, casting information, etc. Similarity computation is done with the help of Pearson Correlation method which is available in the literature. An improved correlation method can be developed to increase accuracy.

References

1. Debnath, S., Ganguly, N., Mitra, P.: Feature weighting in content based recommendation system using social network analysis. In: Proceedings of the 17th International Conference on World Wide Web, pp. 1041–1042 (2008)
2. Marlin, B.: Collaborative filtering: a machine learning perspective. Dissertation, University of Toronto (2004)
3. Melville, P., Mooney, R.J., Nagarajan, R.: Content-boosted collaborative filtering for improved recommendations. In: Proceedings of the 18th National Conference on Artificial Intelligence (AAAI'02), pp. 187–192, Edmonton, Canada (2002)
4. Resnick, P., Iacovou, N., Suchak, M., Bergstrom, P., Riedl, J.T.: GroupLens: An open architecture for collaborative filtering of Netnews. In: Proceedings of the Conference on Computer Supported Cooperative Work. ACM. *xi*+464, pp. 175–186
5. Pazzani, M.J., Billsus, D.: Content-based recommendation systems. In: The Adaptive Web Lecture Notes in Computer Science, vol. 4321, pp. 325–341 (2007)
6. Papagelis, M., Plexousakis, D., Kutsuras, T.: Alleviating the sparsity problem of collaborative filtering using trust inferences. University of Crete, Greece & Forth-ICS (2005)
7. Wen, J., Zhou, W.: An improved item-based collaborative filtering algorithm based on clustering method. In: J. Comput. Inform. Syst. **8**(2), 571–578 (2012)
8. Zhao, Z.-D., Shang, M.-S.: User-based collaborative-filtering recommendation algorithms on Hadoop. In: 2010 Third International Conference on Knowledge Discovery and Data Mining
9. Wang, J., De Vries, A.P., Reinders, M.J.T.: Unifying user-based and item-based collaborative filtering approaches by similarity fusion. In: Proceedings of the 29th Annual International ACM SIGIR Conference on Research and Development in Information Retrieval (2006)
10. Tsunoda, M., Kakimoto, T., Ohsugi, N.: Javawock: A Java class recommender system based on collaborative filtering
11. Sarwar, B., Karypis, G., Konstan, J., Riedl, J.: Item-based collaborative filtering recommendation algorithms. WWW10, Hong Kong, May 1–5 2001. ACM 1-58113-348-0/01/0005
12. Gong, S.: Learning user interest model for content-based filtering in personalized recommendation. Int. J. Digital Content Technol. Appl. (JDCTA) **6**(11) (2012) (June)
13. Content recommendation system based on private dynamic user profile. In: Proceedings of the Sixth International Conference on Machine Learning and Cybernetics, Hong Kong, 19–22 August 2007
14. Balabanovic, M., Shoham, Y.: Fab: content-based, collaborative recommendation (Special section: recommender systems). In: Communications of the ACM, March 1997, vol. 40(3), p 66(7)
15. Xie, F., Chen, Z., Shang, J., Huang, W., Li, J.: Item similarity learning methods for collaborative filtering recommender systems. In: AINA, 2015, 2015 IEEE 29th International Conference on Advanced Information Networking and Applications
16. Chen, Y., Yu, M.: A hybrid collaborative filtering algorithm based on user-item. In: 2010 International Conference on Computational and Information Sciences
17. Hu, L., Song, G., Xie, Z., Zhao, K.: Personalized recommendation algorithm based on preference features. In: Tsinghua Science and Technology. ISSN 1007-0214 08/11, pp. 293–299, vol. 19(3), June (2014)

Classifying Exoplanets as Potentially Habitable Using Machine Learning

Karan Hora

Abstract With the launch of the ESA Gaia satellite observatory and the planned LSST, and the torrent of data coming through the Kepler space observatory, scientists will be able to collect data for more than 1 billion astronomical objects, including millions of exoplanets in the coming years. In this study, several predictive models are built using machine learning algorithms to classify exoplanets as potentially habitable, based on various characteristics of the planet and its star. I applied six supervised learning algorithms for the classification of planets, which include two decision trees, CART and Random Forest, Support Vector Machines, Logistic Regression, Feed-Forward Neural Network, and Naïve Bayes. I further applied CART to create a regression model to predict the value of the ESI (Earth Similarity Index) for an exoplanet.

Keywords Data mining · Decision trees · Random forests · Naïve bayes · Logarithmic regression · Support vector machines · Artificial neural networks

1 Introduction

As our sun, which is 4.6 billion-years old and almost halfway through its lifetime, slowly rises in temperature at its core, scientists suggest that the earth will move out of its habitable zone, and our vast oceans would evaporate in 1.75 billion years. Though not a concern for our generation, it may someday be a problem facing humanity. The Kepler satellite observatory was launched in 2009 to discover Earth-like planets orbiting stars in the Milky Way galaxy. Its goal was to identify planets in the habitable zone of their stars, which is the region around the star wherein a planet with adequate atmospheric pressure can sustain liquid water on its surface. Data from the Kepler space observatory suggests there maybe 40 billion planets similar to Earth orbiting in the habitable zone of their stars in the Milky Way [1, 2].

K. Hora (✉)
Maharaja Surajmal Institute of Technology, New Delhi, India
e-mail: horakaran@outlook.com

A.K. Saini et al. (eds.), *ICT Based Innovations*, Advances in Intelligent Systems and Computing 653, https://doi.org/10.1007/978-981-10-6602-3_20

With the launch of the new Gaia satellite observatory, and the planned Large Synoptic Survey Telescope set to begin operations in 2022, astronomers will be able to see further into space and collect data about billions of astronomical objects, including millions of exoplanets [3]. Hence, a model which could analyze this huge torrent of incoming data to filter out potentially habitable planets based on initial observations would help scientists in their search. In this study, I created several predictive models using different machine learning algorithms to classify exoplanets as potentially habitable based on various characteristics of the planet and its star including the planet's mass, radius, density, gravity, Earth Similarity Index, and others.

2 Literature Survey

Previous works using machine learning, in the scope of exoplanets have been in their detection and confirmation [4, 5].

Transit photometry is a method in which light curves from stars are observed and analyzed. Drops in these light curves are indicative of planets. Though this method results in the highest number of planet detection (94% of exoplanets detected in 2014), it has a high rate of false detection, caused due to the dimming of star's light curve caused by instrument noise and other astronomical factors. Machine learning has been used extensively in this field to make sure the dimming in light curves is caused by planets, and not noise. This is done by training models from light curves of confirmed exoplanets passing in front of their host star, with respect to the position of the observing satellite or observatory, and recognizing the patterns [6, 7].

In this paper, I work on the Habitable Exoplanets Catalog maintained by the Planetary Habitability Laboratory at the University of Puerto Rico at Arecibo [8]. Previous work on this dataset has been only done by Saha et al. [9]. In their study, they applied various algorithms achieving the best predictive accuracy of 98.7%.

3 Dataset

The Habitable Exoplanets Catalog contains 1943 exoplanets observed from various sources, out of which 31 have been classified as potentially habitable. The original dataset contains 68 features, 17 categorical features, and 51 continuous features. Due to the relatively low number of records and a high number of dimensions, the models created from the original data may have overfitted the training set [10] thus leading to slightly lower accuracy in testing. Therefore, the dimensions were pruned to the 28 important features, which gave better results. The selected dimensions, with their units, are listed in Tables 1 and 2. In creating the classification model, I have excluded ESI (Earth Similarity Index) as a training parameter, as it is calculated based on the values of radius, density, escape velocity, and surface temperature which is already being accounted for independently [11].

Table 1 Planet features

S. No.	Dimension	Units
1.	Mass	EU
2.	Radius	EU
3.	Density	EU
4.	Gravity	EU
5.	Escape velocity	EU
6.	Mean equilibrium temperature	K
7.	Minimum surface temperature	K
8.	Mean surface temperature	K
9.	Maximum surface temperature	K
10.	Surface pressure	EU
11.	Planet period	Days
12.	Semi-major axis	AU
13.	Planet eccentricity	
14.	Mean distance from the star	AU
15.	Atmosphere class	None/metals-rich/hydrogen-rich
16.	Habitable zone distance	
17.	Habitable zone composition	
18.	Habitable zone atmosphere	
19.	Habitable zone index	
20.	Standard primary habitability	
21.	Potential habitability	FLAG = 0 or 1

where
EU Earth units
K Kelvin
AU Astronomical units
SU Solar units

Table 2 Star features

S. No.	Dimension	Units
1.	Mass	SU
2.	Radius	SU
3.	Effective temperature	SU
4.	Luminosity	SU
5.	Iron to hydrogen ratio	
6.	Age	Billion years

In addition, I faced the problem of missing data in both continuous and categorical dimensions. I resolved this by imputing the missing value by the weighted average of the non-missing observations, for continuous dimensions, and by imputing the category with the largest average proximity for categorical dimensions.

4 Machine Learning Algorithms

In this paper, I applied six supervised learning algorithms for the classification of planets, which include two decision trees, CART and Random Forest, Logistic Regression, Feed-Forward Neural Networks, Support Vector Machines, and Naïve Bayes.

I further applied CART to create a regression model to predict the value of the ESI (Earth Similarity Index) for a given planet. ESI is a scale from 0 to 1. A planet having a value closer to 1 is very similar to Earth, and closer to 0 indicates very dissimilar to Earth.

5 Experimental Setup

All algorithms are implemented on the R statistical programming languages, using relevant packages. The dataset has been split into training and testing sets, with a ratio of 70:30 for all models except for Support Vector Machines where the dataset is split in the ratio 60:40.

6 Results

In the classification problem, a standard baseline model which would correctly predict the most common output (non-habitable) correctly would give us an accuracy of 98.4%. The aim has been to significantly improve on that score.

The results of the models are represented in the form of a confusion matrix, such as the one below.

Confusion Matrix:

	Predicted class = 0 (non-habitable)	Predicted class = 1 (habitable)
Actual class = 0 (non-habitable)	True negatives	False positives
Actual class = 1 (habitable)	False negatives	True positives

The measures used for quantifying the accuracy of a model based on the outputs of the confusion matrix for each algorithm include:

i. Overall accuracy—the ratio of the number of records in the testing set classified correctly (True Positives and True Negatives) to the total number of records.

ii. Sensitivity—also known as the True Positive rate. It is the ratio of the number of records correctly classified as positive (True Positives), to the actual number of positive records in the test set. (True Positives and False Negatives)
iii. Specificity—also known as the True Negative rate. It is the ratio of the number of records correctly classified as negative (True Negatives), to the actual number of negative records in the test set. (True Negatives and False Positives)

Table 3 illustrates the overall accuracy of all models, compared to the baseline accuracy.

7 CART Model

The Classification And Regression Trees are a technique which can be used to produce either classification or regression trees. This model gives us the best accuracy. In addition, this model does not require a separate algorithm for handling missing data. Figure 1 illustrates the resultant decision tree, where P. Ts Mean represents the planet's mean surface temperature in Kelvin, and P. HZI represents the Planet Habitable Zone Index.

Table 3 Accuracy of all models

Algorithm	Percentage accuracy
CART	99.82818
SVM	99.48454
Random forest	99.31271
Logistic regression	99.14089
Feed-forward neural network	98.79725
Baseline accuracy	98.40452
Naïve Bayes	95.53265

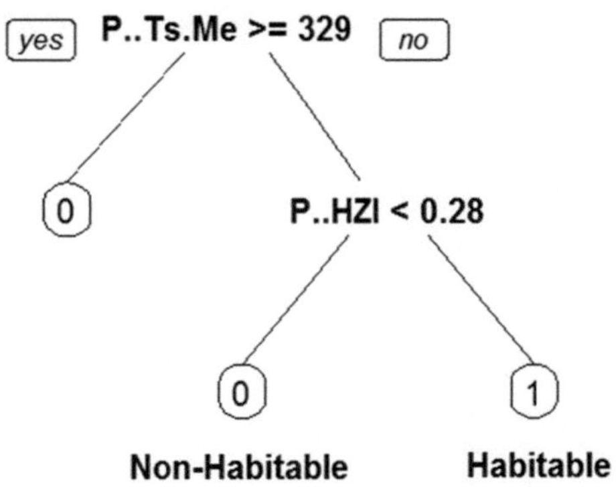

Fig. 1 CART model decision tree

For the CART model, I divided the available dataset into training and testing sets in the ratio of 7:3.

The resultant confusion matrix for the test set is:

	Predicted class = 0 (non-habitable)	Predicted class = 1 (habitable)
Actual class = 0 (non-habitable)	572	1
Actual class = 1 (habitable)	0	9

Overall accuracy = 99.82818
Sensitivity = 1
Specificity = 0.998

Figure 2 represents the regression tree, which here has been trained to predict the value of ESI (Earth Similarity Index) for an exoplanet. The value of the ESI is hugely beneficial in interpreting the potential habitability of the planet. Planets with ESI values between 0.8 and 1.0 have a high chance of being habitable.

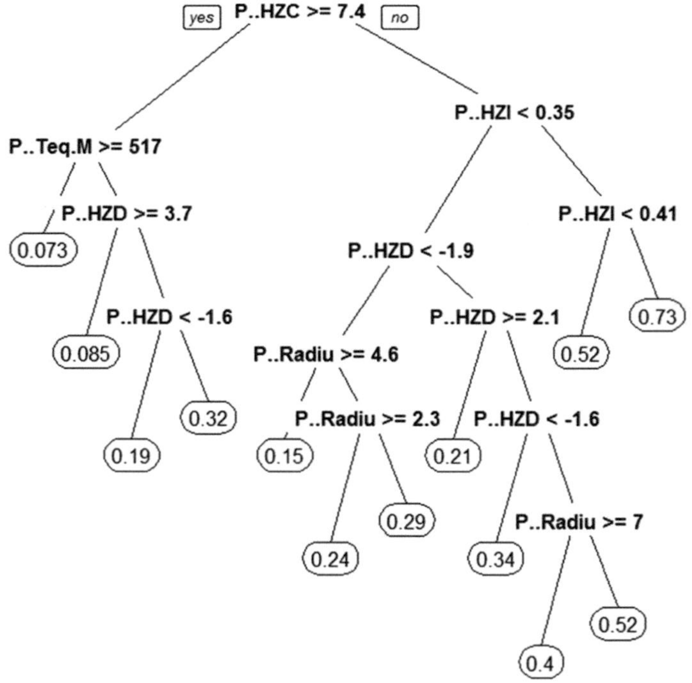

Fig. 2 CART model regression tree predicting ESI (Earth Similarity Index)

8 Support Vector Machines

SVMs are non-probabilistic binary linear classifiers, which build models which map points into the sample space such that different categories are separated by a visual distance. For the SVM model, I divided the available dataset into training and testing sets in the ratio of 6:4.

The resultant confusion matrix for the test set is:

	Predicted class = 0 (non-habitable)	Predicted class = 1 (habitable)
Actual class = 0	764	4
Actual class = 1	0	8

Overall accuracy = 99.48454
Sensitivity = 1
Specificity = 0.994

9 Random Forests

Random Forest is the ensemble learning variant of decision trees wherein 'n' numbers of trees are created, randomly selecting features and then creating a single model from those trees. For the Random Forests model, I divided the available dataset into training and testing sets in the ratio of 7:3.

The resultant confusion matrix for the test set is:

	Predicted class = 0 (non-habitable)	Predicted class = 1 (habitable)
Actual class = 0 (non-habitable)	569	1
Actual class = 1 (habitable)	0	9

Overall accuracy = 99.31271
Sensitivity = 1
Specificity = 0.998

10 Logistic Regression

It is a regression model which here has been used to classify the dependent variable into two categories, habitable or non-habitable. For the Logistic Regression model, I divided the available dataset into training and testing sets in the ratio of 7:3.

The resultant confusion matrix for the test set is:

	Predicted class = 0 (non-habitable)	Predicted class = 1 (habitable)
Actual class = 0 (non-habitable)	568	5
Actual class = 1 (habitable)	0	9

Overall accuracy = 99.14089
Sensitivity = 1
Specificity = 0.991

11 Feed-Forward Neural Network

A feed-forward neural network is an artificial neural network where the connections always move in the forward direction, from the input nodes to the output nodes, and never form a directed cycle. For the feed-forward neural network model, I divided the available dataset into training and testing sets in the ratio of 7:3.

The resultant confusion matrix for the test set is:

	Predicted class = 0 (non-habitable)	Predicted class = 1 (habitable)
Actual class = 0 (non-habitable)	566	0
Actual class = 1 (habitable)	7	9

Overall accuracy = 98.79725
Sensitivity = 0.562
Specificity = 1

12 Naïve Bayes

It is a probabilistic classifier based on applying Bayes' theorem with strong independence assumptions between the features, which are used to calculate the posterior probabilities. For the Naïve Bayes model, I divided the available dataset into training and testing sets in the ratio of 7:3.

The resultant confusion matrix for the test set is:

	Predicted class = 0 (non-habitable)	Predicted class = 1 (habitable)
Actual class = 0	547	0
Actual class = 1	26	9

Overall accuracy = 95.53265
Sensitivity = 0.25
Specificity = 1

13 Conclusions

By using various supervised learning classifications algorithms such as CART, Random Forest, Support Vector Machines, Logistic Regression, Feed-Forward Neural Network, and Naïve Bayes models were developed to predict the potential habitability of recently observed exoplanets. All models, apart from Naïve Bayes, performed better than the baseline accuracy, with the CART model performing the best with an accuracy of 99.89%. Further, the models can allow astronomers and scientists to draw insights from the data and help understand which characteristics carry greater weightage in deciding habitability. In addition, I also created a regression tree to predict the value of Earth Similarity Index for a planet (ESI). By analyzing the tree one can interpret which factors, apart from the ones used to calculate its value, can affect the value of ESI, which was hitherto unexplored.

The exoplanet dataset has grown from only a handful of planets in the 1990s, to today expanding almost on a daily basis through observations of the Kepler satellite and other sources [12]. Hence there has been a great interest in finding potentially habitable planets with the exoplanets Kepler-438b, which has the highest value of ESI (0.88) of any exoplanet observed by man, and the Kepler-452b with ESI of 0.83, both were discovered in 2015. Therefore, these models can be further optimized and be used to sift through the deluge of incoming data to pinpoint potentially habitable planets, which can then be thoroughly analyzed by astronomers and scientists.

References

1. Borucki, William J., et al.: Kepler planet-detection mission: introduction and first results. Science. **327**(5968), 977–980 (2010)
2. Jenkins, J., Smith, J., Peter Tenenbaum, J.T, Van Cleve, J.: Planet Detection: The Kepler Mission (2014)
3. Lindegren, Lennart: Gaia: Astrometric performance and current status of the project. Proc. Int. Astron. Union **5**(S261), 296–305 (2009)
4. Way, M.J., et al. (eds.): Advances in Machine Learning and Data Mining for Astronomy. CRC Press, Boca Raton (2012)
5. Bloom, J., Richards, J.: Data mining and machine-learning in time-domain discovery & classification. Adv. Mach. Learn. Data Min. Astron. (2011)
6. Debosscher, J., Aerts, C., Vandenbussche, B.: Classification of COROT exoplanet light curves. Astrophys. Variable Stars. **349**, (2006)
7. McCauliff, Sean D., et al.: Automatic classification of Kepler planetary transit candidates. Astrophys. J. **806**(1), 6 (2015)
8. Planetary Habitability Laboratory at the University of Puerto Rico at Arecibo, http://phl.upr.edu/
9. Saha, S., Agrawal, S., Manikandan, R., Bora, K., Routh, S., Narasimhamurthy, A.: ASTROMLSKIT: A New Statistical Machine Learning Toolkit: A Platform for Data Analytics in Astronomy. arXiv:1504.07865 [cs.CE] (April 2015)

10. Smith, M.R., Martinez, T.: Improving classification accuracy by identifying and removing instances that should be misclassified. In: The 2011 International Joint Conference on Neural Networks (IJCNN), pp. 2690–2697. (2011)
11. Schulze-Makuch, D., Méndez, A., Fairén, A.G., von Paris, P., Turse, C., Boyer, G., Davila, A.F., António, M.R.D.S., Irwin, L.N., Catling, D.: A Two-Tiered Approach to Assess the Habitability of Exoplanet (2011)
12. Petigura, E.A., Howard, A.W., Marcy, G.W.: Prevalence of Earth-size planets orbiting Sun-like stars. In: Proceedings of the National Academy of Sciences of the United States of America. arXiv: (October 31, 2013)

Model for Detecting Fake or Spam Reviews

Manisha Singh, Lokesh Kumar and Sapna Sinha

Abstract Internet has opened door for new style of marketing. People are using Internet for getting reviews and making opinion about products, movies, or services. Online reviews play important role in electronic commerce, but at the same time, it is also evident that online reviews are not authentic because fake reviews are formulated to gain competitive advantage by the competitors. Spammers write fake reviews to promote or demote target products which deceive customers. In this paper, authors have designed a model to detect fake review and spams along with blocking of such reviewers.

Keywords Opinion mining · Fake review

1 Introduction

Product reviews play a keen role in providing the information to consumers, retailers, and manufacturer which influence their decisions capabilities of purchasing products. These reviews not only help the consumers to get information of products but also provide their feedback for the product they have used. These information are used and important features are added back into the marketing. In paper (Cone 2011), the author has illustrated that approximately 80% of the consumers changes their decision after reading reviews given by existing customers and if the reviews are positive 87% people purchase that product.

With the large development and growth of Internet, it has become a trend that people rely a lot on online reviews to get the information of products and on the

M. Singh (✉) · L. Kumar (✉) · S. Sinha
Department of IT, ASET, Amity University, Noida, India
e-mail: sweetmani2592@gmail.com

L. Kumar
e-mail: lokesh.skshukla@gmail.com

S. Sinha
e-mail: ssinha4@amity.edu

A.K. Saini et al. (eds.), *ICT Based Innovations*, Advances in Intelligent Systems and Computing 653, https://doi.org/10.1007/978-981-10-6602-3_21

basis of that they make their decisions to purchase the product. But this also leads to the issues of spam reviews whose main intentions are to mislead the readers by giving positive or negative opinions.

Most of the reviewers write only one review (called Singleton review) [guan wang] but due to the nature of spamming a spammer too many reviews in a short duration under various names. In this paper, we provide the following contributions. First, we classify the reviews into three categories based on their rating, one is bad (1 and 2 star), second is neutral (3 star), and third is good (4 and 5 star). Second, we apply different assessment methods.

2 Literature Survey

Most of the work that has been done is mainly focussed on spam detection in e-mails. However, there are very few studies that have been focused on fake review detection. In (Liu and Jindal), authors use the reviewer's profile, product details, and review text to detect the duplicate and near duplicate opinions. They also made use of duplicate or nonduplicate opinion which explains the deceptive opinion and deceptive opinion, respectively. A deceptive opinion will have similar text on the same website whereas the nondeceptive opinion will be considered as truthful. Although duplicate opinions can be somewhat detected with plagiarism detection tools.

In (Wu et al. 2008) focusses on distortion of posting patterns based on popularity rankings. Their study was based on the hotel reviews that have been received and on the profile of the reviewers, whereas we are focussing on the point whether the same person is posting multiple reviews for the same product or brand.

Zhang and Paxon focus on detecting automated activity on twitter. According to them, if an account is automated it will follow certain timing patterns. That is the timing patterns will be too uniform or nonuniform at all.

In (Rakesh Verma and Ekene Sibeudu), authors use feature selection as a mechanism to detect opinion spam for Twitter. The author tries to identify that whether the tweet which is posted on the twitter is either posted by a bot or human. To achieve, they build a classifier based on features.

Some of the recent studies also examine the use of human computation and unexpected rules in detecting fake or spam reviews.

3 Proposed Model

Spammers tend to post a large number of reviews (highly or low rated) to promote or demote a product or service. A review can have certain attributes, e.g., user_id, user_ip_address, brand_id, product_id, product_rating and product_review, as many of the major websites track their user's IP addresses. Hence, we have incorporated IP address as an attribute (user_ip_address) along with the above-mentioned attributes. Based on the product_rating given by the reviewer, our system classifies the review into one of the following categories—bad, neutral, or good.

The whole process is divided into three steps as shown in Fig. 1a, b, c.

1. Initially, the reviewer will post a review regarding a product or services. The review goes through the library and tries to look for existing matching records. The verification takes places by searching the same details in the library that has been maintained. The whole library is searched to check whether the corresponding detail of user is present or not. Now the results can either match or not. If user_id and product_id matches. Generate error message that you cannot rate same product twice. If it not matches then it will try to look up for other conditions such as: user created a new user_id in order to rate same product again, user is rating different products of same brand (promoting or demoting a certain brand), etc.

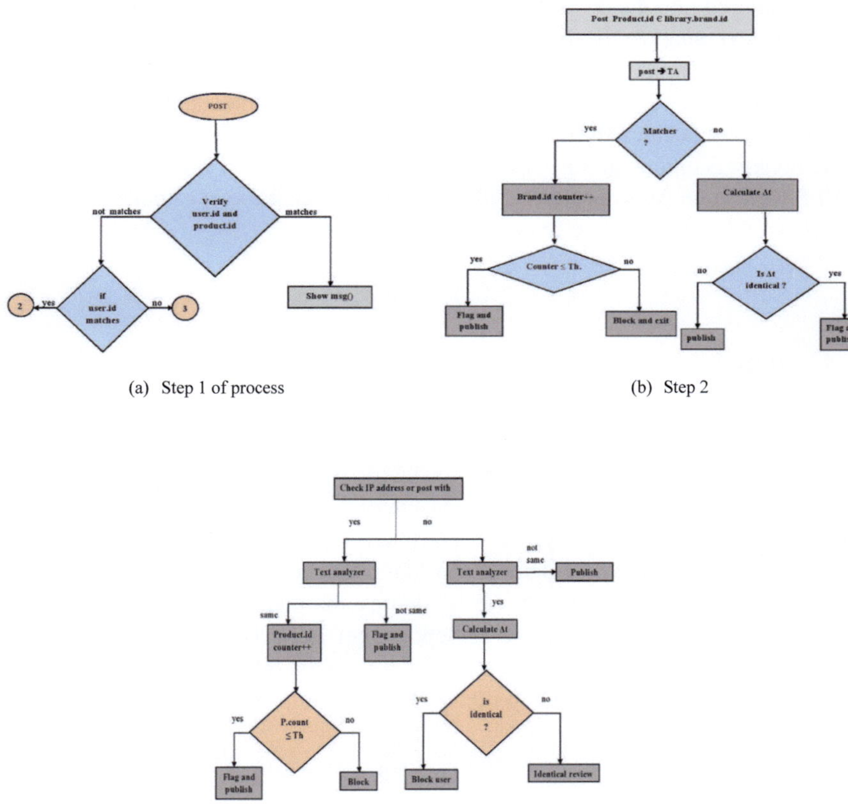

(a) Step 1 of process

(b) Step 2

(c) Step 3 of the process

Fig. 1 a, b, c The process of detecting fake review

2. If the user_id gets matched then it means that the user had given the review in past. Again we come across two results, either it matches or not. If it matches then the product.id is matched from library.brand.id that means whether it belongs to the same brand or not and we post this detail on the text_analyzer (TA). The text analyzer will give the result accordingly. If the result matches that means the product is of the same brand and in this case, we will consider a brand.id counter that will ensure that how much times the reviewer is posting the reviewing on the same brand product. So for this, a threshold will be taken that will decide the nature of the review. If the condition counter <= threshold (th) is true then the review is posted but with a mark that will indicate that it may be a spam review and if the condition of threshold is not true then we will block the user and exit. But, if the text analyzer result comes out to be NO then in that case, we are going to check the difference between the timestamp associated to the two consecutive posts. This will be like $t1 - t2 = \Delta t$. If it comes out to be identical then the review will be posted with the mark indicating that it may be a spam review. And if it is not identical then post the review.
3. Once entered into the third step, check the ip_address of the user and also it gets into the text_analyzer. If the ip_address is same and the result of text_analyzer is same then the product.id counter is taken whose threshold condition decides to block the user or to publish the post with mark. But is if not same then is not the text analyzer decides further results. If the result shows that it is not same then the review is unique and it will be published. On the other hand, if the result is YES then again Δt is checked. If identical then block the user else it will post the review and will show the review is identical.

4 Conclusion and Future Scope

Detecting fake or spam review is an important activity in opinion mining. As fake review can mislead a consumer to buy a faulty product whereas there are many other better options are also available. As internet marketing has been grown significantly during the previous years. Fake review can also be used as a medium to target brands, i.e, promote or demote the brands or products of these brands. We have proposed a model which can easily identify the fake reviews (might be fake) and fake reviewers and block the review to get published and also block the reviewer to post any further fake reviews. As people are relying more on reviews to make their decisions about the product. We can implement this model into a real-time service.

References

1. Mukherjee, A., Liu, B., Wang, J., Glance, N., Jindal, N.: Detecting group review spam. WWW'11
2. Jindal, n., Liu, B., Lim, E-P.: Finding unusual review patterns using unexpected rules. CIKM'10
3. Bond, C.F., DePaulo, B.M.: Accuracy of deception judgements. Pers. Soc. Psychol. Rev. **10**(3), 214 (2006)
4. Mukherjee, A., Liu, B., Glance, N.: Spotting fake reviewer groups in consumer reviews. In: Proceedings of International World Web Conference (WWW-2012), (2012)
5. Liu, B.: Sentiment Analysis and Opinion Mining, Morgan & Claypool Publishers, San Rafael (May 2012)
6. Fusilier, D.H., Guzm´an Cabrera, R., Montes-y-G´omez, M., Rosso, P., Using PU-learning to detect deceptive opinion spam. In: Proceedings of the 4th Workshop on Computational Approaches to Subjectivity, Sentiment and Social Media Analysis, pp. 38–45. Atlanta, Georgia (14 June 2013)
7. Prajapati, J., Bhatt, M., Prajapati, D.J.: Detection and summarization of genuine review using visual data mining. Int. J. Comput. Appl. (0975–8887). **43**(11) (April 2012)
8. Verma, R., Sibeudu, E.: Robust and effective selection for opinion spam detection. In: 28th Annual Computer Security Applications Conference (ACSAC). Orlandlo, FL (December 2012)

Towards Understanding Preference of Use of Emoticons for Effective Online Communication and Promotion: A Study of National Capital Region of Delhi, India

Anil Kr. Saini, Puja Khatri and Khushboo Raina

Abstract With introduction of new tools and techniques for online communication, the methods of online marketing have also changed. The presence of emotions in decision-making processes of customers has always changed the dimensions of sales in online environment. Emoticons, which are the graphical representation of one's nonverbal traits, help in analysing the meaning and emotion of the message in an appropriate manner. Now, increasing usage of emoticons in textual messages by gen y students has revived the emotional aspect of making purchases. These youngsters form a majority portion of online shoppers and always remain the favourite segment of marketers. Addressing their needs by capturing their emotional side at a correct instance will convert the search into buying decision. The present study attempts to understand the perceptions of these students towards the usage of emoticons for effective online promotion. The study was conducted in the National Capital Region of Delhi, India, wherein the respondents were selected through multistage sampling ($N = 120$). Data was collected through self-constructed questionnaire (Cronbach alpha = 0.872). Differences between male female respondents as regards their perception of usage of emoticons were also probed for gaining insights into the emotional dimensions of both the groups while making online purchase. The study will be a good source for online marketers to develop strategies to attract this most vivacious segment by making use of emoticons.

Keywords Emoticons and emotions · Emoticons in online promotion · Computer mediated communication · Emoticons and gen y · Emoticons and online advertising · Emoticons usage in AIDA

A.Kr. Saini (✉) · P. Khatri · K. Raina
University School of Management Studies, Guru Gobind Singh Indraprastha University, Delhi, India
e-mail: aksaini1960@gmail.com

P. Khatri
e-mail: pujakhatri12@gmail.com

K. Raina
e-mail: khushboo0803@gmail.com

A.K. Saini et al. (eds.), *ICT Based Innovations*, Advances in Intelligent Systems and Computing 653, https://doi.org/10.1007/978-981-10-6602-3_22

1 Introduction

Humans, since birth, are capable of communicating with other humans. Observing facial expressions and understanding nonverbal cues, individuals can understand the reaction of other individual without speaking. With advent of technology, the communication channels have been greatly modified. It is an agreed fact that technology has revolutionized the way we live. Smart phones allow us to communicate, work, make online purchases and what not in the easiest manner; lives have been made easier especially in terms of staying connected. Out of so many tools and techniques to communicate over technology enabled devices, 'instant messaging' has become the most popular means to communicate socially and companies are implementing this technology to make the workplace more collaborative and finding business opportunities [16]. Despite the advantages, online communication lacks the most powerful nonverbal cues like facial expression and gestures which are crucial for expressing attitude and opinions; introduction of emoticons has helped in overcoming this void to great extent [7]. The ability to show emotions in messages to express one's emotions is very important for a socially amiable environment which further facilitates friendship [26]. Communication through IM can be enriched by the creation and usage of emoticons and emotions can be expressed freely by means of emoticons while communicating online. Walther [24] opines that use of emoticons in messages allows users to enhance their decoding skills to derive textual cues in order to form interpersonal impressions. Emoticons are taken as indicators of emotional states whose purpose is to convey that nonverbal information which otherwise are conveyed through facial expressions and body language [9]. Emoticons have been accepted globally as an indicator of nonverbal communication in writing. Companies have started harnessing this concept to attract new generation buyers over online portals as youth is the largest segment to make purchases online. 89% of the youth in India research for things online before making purchase [8]. Their perception and way at looking things matter, if their intentions have to be converted into final buying decisions.

The present study attempts to analyse the preference of making use of emoticons for effective online promotions. The study maps the perceptions of generation y online buyers as regards their awareness, interest, desire, and actions, if emoticons have been used to bring them to decisions to make purchases online. The researchers were not able to find any published work in this direction and especially in context of India. Therefore, this research forms a base for further studies in the same direction and provides an insight to online retailers and marketers to develop strategies to attract this.

2 Literature Review

Human emotions form a significant part of our lives and play very important role in understanding the behaviour of user while interacting with computer [6]. Understanding of human emotions determines the consumer purchase behaviour and its influence on their decision making. According to Zajonc [27], emotions play an indispensable role in marketing activities and are capable to override cognitive abilities of humans. In a study conducted by McDuff et al. [18], emotions are best expressed by facial expressions and by coding their facial expressions, they established its linkage with advertisement driven sales effectiveness. Bindu et al. [4] developed a model of recognizing emotions from their facial expressions and thus classified 22 emotions: Happy, Pride, Enthusiasm, Joy, Love, Tenderness, Ecstasy, Lust, Surprise, Conformity, Boredom, Indifference, Disgust, Fear, Revenge, Rage, Sadness, Hate, Grief, Shame, Sorrow, and Anger. Emotions in textual forms are best expressed by the use of emoticons. According to Dresner and Herring [9], there can be three ways in which emoticons can function: (a) as indicators of emotion which can be directly mapped onto facial expression, (b) as nonemotional indicators which are mapped conventionally onto facial expressions and (c) as nonverbal forced indicators which do not conventionally get mapped onto facial expressions. Park [20] defines emoticons as '*graphical representations of interpersonal and emotional features, expressed through gesture and facial expressions in face-to-face interactions, in the online setting*' (pp. 151). Emoticons represent body language in the form of text while communicating over messages and where the communication channel only transmits alphabets and punctuations (Fig. 1).

The history of usage of emoticons is still ambiguous but for the first time the typographical emoticons for a class of joy or melancholy were appeared in the U.S. magazine 'Puck' published in 1881 [21]. Emoticons are widely used in online communication these days but their usage is restricted to text messaging. The

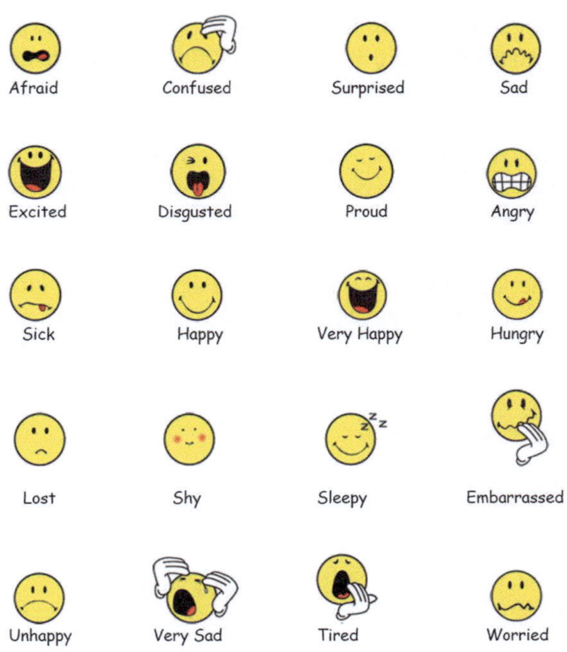

Fig. 1 Image of few emoticons with meaning

organizations dealing through online portals can make best use of this feature of textual communication to attract the desired segment as emoticons can express their non spoken thoughts. Also, emoticons may have greater influence while communicating with the customers in online environment. The advertisements to lure the young segment have to be designed in some creative manner so as to catch their attention immediately. Several models have been framed for discussion of effectiveness of advertising; the most prominent model used by researchers in this context is that of developed by Elmo Lewis in 1906, which is known as AIDA model. It identifies the different cognitive stages through which an individual goes during purchase process, here, A stands for Awareness: creating brand awareness, I for Interest: generate interest in the client towards products, D for Desire: create enthusiasm or stimulate people for purchase and A for Action: when decision has been made and the process terminates with buying [10] (Fig. 2).

The usage of emoticons depends on one's personality and gender as well. Ho and Vathanophas [13] opine that personality characteristics affect the process of online discussion and the outcome of this discussion. On the basis of this notion, Xu et al. [26], put forth that interpersonal perceptions influence communication which means that perception of sender may vary according to the receiver's personality. According to Herring [12], since inception of computer networks, men have dominated the technology and formed a major portion of its users. This is because males follow more logical and technical cues than females. Females have said to be more emotionally expressive than males. Males tend to restrict their certain emotions when it comes to communication. According to Kelly and Hutson-Comeaux [17], emotions of joy, sadness and fear are integral traits of females whereas aggression is the characteristic of males. A number of studies have been conducted on the differences in emotional expression of both the groups, but in context of perception towards online promotions, researchers were not able to locate any study in Indian context.

As, the emotional expression of both the groups is different, the significance of emoticons for their communication purpose also varies. Wolf [25] conducted a study to probe gender differences in use of emoticons on newsgroup and expressed the term 'emoticon' as 'emotional-icon'. The author found that females make use of emoticons more to express humour rather than sarcasm whereas men used emoticons to show sarcasm more. Though, both the groups take liberty to change the default definitions and meanings of emoticons to express their emotions. Females have enriched the male definitions and use of emoticons by adding dimensions like

Fig. 2 AIDA model

gratitude, support, assertion of positive feelings. Herring [11] conducted study to confirm that females use more emoticons in their textual messages than males for they are more emotionally expressive than them. In another study conducted by Hwang [15] to ascertain the gender differences among college students of a university in Seoul, it was found that there is a difference in the usage as well as motivation in using emoticons in messages. Females use emoticons more to express intimacy and give meaning to their messages as compared to males as emoticons act like socioeconomic support to them. Though studies have been found, probing the gender differences in emoticon use but in context of online promotions, studies are missing specially in Indian relevance. This study will analyse the gender differences in emoticon usage in the context of online promotions so that the most booming sector in terms of advertisements and sales can gain some relevant and useful insights.

3 Objectives

1. To study the significance of emoticon usage in communication.
2. To study the perception of the respondents regarding their awareness, interest, desire and action generated in message using emoticons.
3. To study the gender differences in level of preference of emoticons usage in messages used for online promotion.
4. To study the gender differences in perceptions of respondents regarding preference given to promotional message which uses emoticons.
5. To probe the gender differences in perceptions of respondents regarding capability of emoticons to attract more customers.
6. To give recommendations to online marketers regarding usage of emoticons for online promotions in order to target the young segment of online shoppers.

4 Hypotheses

Ho1 There exists no difference between male and female respondents regarding their perception towards level of awareness created by emoticons for a product.
Ho2 There exists no difference between male and female respondents regarding their online research for products on the basis of emoticons used.
H3 There exists a difference between male and female respondents regarding their perception towards capability of emoticons to attract more customers.
H4 There exists a difference between male and female respondents regarding preference given to a promotional message containing emoticons.

5 Research Methodology

This research initiative studies the perceptions of Management and IT students in Delhi-NCR as regards the various items of AIDA model in context of usage of emoticons in online promotions. The research has been carried out with a self-constructed questionnaire. The questionnaire had several items related to the perception of respondents regarding awareness, interest, desire and action taken by respondents when they see emoticons in online promotions. The questionnaire had two parts, part A was purely focused on collecting demographic details and part B mapped the perceptions of students regarding different items of the AIDA model and gender differences were probed for the same set of items. The questionnaire was constructed on a five-point Likert agreement scale to measure the responses on the decided variables. The questionnaire was subjected to review by experts and their inputs have been incorporated accordingly. Reliability of the same was computed to be Cronbach Alpha (0.872). According to Nunnally (1978, p. 245), the instruments used in basic research have a reliability of about 0.70 or better.

The sampling was multistage. In the first place, it was purposive wherein the researchers drew out a list of management and technical institutions affiliated by AICTE (All India Council for Technical Education). The list was generated through the web link http://trueguides.blogspot.in/2011/03/list-of-aicte-approved-colleges-in.html. At second stage, sampling was stratified wherein self-financed management and technical institutes were selected from the list. Out of it, two management and two technical institutes were selected for the study. Final year students were administered questionnaires so as to record their perceptual inferences regarding various items of a total of 40 students per institute filled the questionnaires; out of 160 filled questionnaires 120 questionnaires were taken for the study.

6 Data Analysis

The analysis of the Table 1 which maps the perception of gen y students as regards their 'Awareness' of usage of emoticons in online promotions, reveals that majority of the students believe that emoticons are capable of creating a great influence while visiting online portals (m = 3.658, s.d. = 0.855). This is because the purchase behaviour of youth is driven by their emotions. They take certain buying decisions

Table 1 Perceptions of respondents as regards their 'Awareness' of usage of emoticons in online promotions

S. No.	Items	N	Mean	Std deviation
1	Influence created by emoticons	120	3.658	0.855
2	Emoticons attract more customers	120	3.492	0.81
3	Awareness created by emoticons as regards the product	120	3.225	0.884

owing to what emotion they contain at that instant and preference to focus on a particular brand is also driven by emotions [19]. The presence of emoticons in message fosters that particular emotion and the message/advertisement sound influential to them. This forms the basis of the strong perception put forth by many respondents that emoticons are capable of attracting customers ($m = 3.492$, s.d. = 0.81). Emoticons are the written form of nonverbal gestures which you give while communicating online [2] and therefore, the presence of an emoticon conveys some meaning with respect to the type of product or its usage. This is the reason that many respondents believe that emoticons facilitate awareness as regards the particular product ($m = 3.225$, s.d. = 0.884).

Table 2 focuses on the ‘Interest’ generated to buy the product by use of emoticons in online promotions. The analysis shows that gen y gives preference to the messages containing emoticons ($m = 3.717$, s.d. = 0.758), this happens because the youth or young population is the major user of emoticons in daily textual messaging. They try to fill the void created by non-face-to-face meetings through these emoticons and relate them to the emotion felt by other party. According to Vision Critical [14], to understand the millennial’s for gaining insights into their buying behaviour, one needs to put on their shoes to get there. They capitalize on social media more than other avenues to make search and decisions. The youth makes use of Internet for research of their favourite brands. The use of emoticons is also favoured by the millennials only worldwide. So, many of the respondents feel that use of emoticons can keep customers engaged ($m = 3.283$, s.d. = 0.832) and the online research for products and information is also encouraged by use of emoticons ($m = 3.258$, s.d. = 0.884). Emoticons are the replacement of facial expressions in written form when face-to-face meetings are not happening. Mere presence of emoticons cannot facilitate decision to buy some product online. Thus, not many respondents feel that emoticons are enough to convey information regarding a product ($m = 2.883$, s.d. = 0.972).

Table 3 shows the perceptions of respondents as regards their ‘Desire’ created to purchase a product by use of emoticons in online promotions. According to Seiter [22], different set of emotions represents different wants of the human beings, like happiness makes one want to share, sadness promotes desire to connect and empathize, fear makes one to stick to something and anger shows stubbornness. By use of similar emoticons one can make the customer enquire about it more as that

Table 2 Perceptions of respondents as regards their ‘Interest’ so generated to buy the product by use of emoticons in online promotions

S. No.	Items	N	Mean	Std deviation
1	Preference to a message containing emoticons	120	3.717	0.758
2	Usage of emoticons to engage customers	120	3.283	0.832
3	Online research for products can be encouraged by emoticons	120	3.258	0.884
4	Information conveyed by emoticons	120	2.883	0.972

Table 3 Perceptions of respondents as regards their 'Desire' created to purchase a product by use of emoticons in online promotions

S. No.	Questions	N	Mean	Std deviation
1	Emoticons enhance enquiry for a product	120	3.567	0.742
2	Desire created by emoticons to purchase the product	120	3.542	0.969
3	Customers prefer to buy product which uses emoticons in their promotion	120	3.467	0.849

want/desire will encourage further research, this may be the reason that majority of the students feel that emoticons are capable of enhancing enquiry for a product (m = 3.567, s.d. = 0.742). When the appropriate emotion is captured by the marketer, the enquiry can be converted into the desire to purchase the product (m = 3.542, s.d. = 0.969). Emoticons are a powerful means of online communications which addresses the emotional need of a human being, therefore, many respondents believe that use of emoticons can actually alter someone's preference to buy that product (m = 3.467, s.d. = 0.849).

The analysis of the perceptions of respondents as regards their 'Action' to make purchase of products promoted online and making use of emoticons are shown in Table 4. The investigation suggests that majority of the respondents agree that emoticons used in online promotions can make customer to try the product (m = 3.392, s.d. = 0.873). This happens because emotions which are the prime reason for aberrations in buying behaviour are rightly countered by the marketers as they understand the need of the hour. The customers driven by emotions see emoticons and try to capture the tacit information so given by the marketer and thus, many of them believe that accurate usage of emoticons can influence customers to express their intentions towards purchase (m = 3.317, s.d. = 0.889). The analysis indicates that a strong and positive view exists among the online buyers as regards the usage of emoticons in online promotions.

For the purpose of testing hypotheses, Levene's t-test for equality of variances (Table 5) has been applied. It has been found that there is no difference between male and female respondents as regards their perception towards level of awareness created by emoticons for a product ($t = -0.435$, $p < 0.05$). This is because both the groups understand the significance of emotionally linked emoticons in their messages. There is no denying to the fact that generation y students are tech savvy and

Table 4 Perceptions of respondents as regards their 'Action' to make purchase of products promoted online and making use of emoticons

S. No.	Questions	N	Mean	Std deviation
1	Emoticons can drive a customer to try the product	120	3.392	0.873
2	Use of emoticons influence customers to express their intention to buy	120	3.317	0.889

Table 5 Differences in the perceptions of male and female respondents as regards the usage of emoticons in online promotional messages

Independent samples test							
		Levene's test for equality of variances		*t*-test for equality of means		
		F	Sig.	*T*	df	Sig. (2-tailed)
Awareness created by emoticons as regards the product	Equal variances assumed	2.423	0.122	−0.435	118	0.665
	Equal variances not assumed			−0.432	112.327	0.667
Emoticons attract more customers	Equal variances assumed	0.182	0.671	−0.67	118	0.504
	Equal variances not assumed			−0.667	114.116	0.506
Influence created by emoticons	Equal variances assumed	0.282	0.596	−0.956	118	0.341
	Equal variances not assumed			−0.956	116.762	0.341
Preference to a message containing emoticons	Equal variances assumed	1.055	0.306	−1.245	118	0.216
	Equal variances not assumed			−1.247	117.476	0.215
Information conveyed by emoticons	Equal variances assumed	0.935	0.336	0.253	118	0.801
	Equal variances not assumed			0.255	117.688	0.799
Usage of emoticons to engage customers	Equal variances assumed	0.039	0.843	0.033	118	0.974
	Equal variances not assumed			0.033	117.542	0.974
Online research for products can be encouraged by emoticons	Equal variances assumed	0.145	0.704	−0.469	118	0.64
	Equal variances not assumed			−0.47	117.644	0.639

keep themselves abreast with the new technology and their tools. They understand that the inclusion of emoticons in the messages aid recipients in having better understanding and gauging the direction of emotions so expressed in the messages in online environment [23]. Therefore, the $H_0$1 stands accepted. The influence generated by emoticons on their audience is equal. The consumers do not look for products which meet needs and rational functions but which become a repository of symbolic meaning, source of relationships, feelings and emotions [5]. The research for the information of products online is done by both the groups equally. The presence of retailers on Internet enables them to look out for other options and performs price comparisons to arrive at certain buying decision [1]. The research for products keeping in mind the cost, convenience, delivery time and security of personal information affects both the group. Hence, the null hypothesis Ho2 which says that there exists no difference between male and female respondents regarding their online research for products on the basis of emoticons used stands accepted. By virtue of being more emotional than males, females tend to make use of emoticons more than males and get more influenced by their use in messages and the context of usage differs [3, 25]. According to studies, there is a difference between both the groups as regards their level of attraction towards use of emoticons but the analysis reveals that there exists no difference between these two groups of respondents for the said variable ($t = -0.670$, $p < 0.05$). This happens because with the increased usage of emotionally linked emoticons in the textual messages, the young men have also started thinking that such usage addresses the emotional side much better than the plain text and can lure the desired segment towards a particular promotional message. Hence, H3 which states that there exists a difference between male and female respondents regarding their perception towards capability of emoticons to attract more customers stands rejected. According to Marcia St. Pierre, a Northern Alberta clinical counsellor [3], emoticons appeal to women more than men because they act as linguistic softener and generally in social settings women tend to weaken their opinions in order to win acceptance and thus make use of emoticons for this purpose. However, the investigation reveals that both groups give more preference to messages which contain emoticons and therefore they think that such promotional messages are preferred by youngsters when they look for making some online purchases ($t = -1.245$, $p < 0.05$). Owing to such great preference given by youth to animated expressions, Facebook has also introduced features of smileys and latest stickers so that emotions can be expressed freely. Therefore, the H4 which says that there exist a difference between male and female respondents regarding preference given to a promotional message containing emoticons stands refuted (Table 6).

Table 6 Group statistics

	Gender	N	Mean	Std. deviation	Std. error mean
Awareness created by emoticons as regards the product	Male	63	3.191	0.85868	0.1082
	Female	57	3.263	0.97333	0.1289
Emoticons attract more customers	Male	63	3.444	0.77829	0.0981
	Female	57	3.544	0.84664	0.1121
Influence created by emoticons	Male	63	3.587	0.85449	0.1077
	Female	57	3.737	0.8562	0.1134
Preference to a message containing emoticons	Male	63	3.635	0.76836	0.0968
	Female	57	3.807	0.74255	0.0984
Information conveyed by emoticons	Male	63	2.905	1.04286	0.1314
	Female	57	2.86	0.8952	0.1186
Usage of emoticons to engage customers	Male	63	3.286	0.85059	0.1072
	Female	57	3.281	0.81841	0.1084
Online research for products can be encouraged by emoticons	Male	63	3.222	0.90597	0.1141
	Female	57	3.298	0.8653	0.1146

7 Conclusion and Implications

Brands are the supplying centres of emotional outflows which results in better relationships with the customers with the ability of inspiring them to tell stories that creates enthusiasm with integration of communication, quality, tradition and identity [5]. The emotional values are linked with person's decision-making. This bias has always been there and in almost all sectors like stock market, etc. This emotional tendency can make customers change their opinions very quickly. The present human race is harnessing the utmost advantages of the information technology, as there are many communication tools which are used to bring people to talk in online environment. Making use of such cost effective means, online retailers makes maximum use of these tools. Now, this emotional part has also been captured by the marketers y introduction of emoticons which are the graphical representation of one's emotional traits. The linkage of one's emotions with some textual message is indeed a revolution. Youth, which forms the major segment of online shopping sector, makes frequent use of such messages in their daily textual messages. The study so conducted reveals that the effect of emoticons is similar on both the groups and both males and females believe that emoticons have the capacity of addressing one's emotions correctly in textual messages. A meaning is linked to every emoticon which represents the state of one's emotion and thus some information can be captured about the product. They create awareness, are influential, promote enquiry about the products, create desire to want the product and thus can make them buy the product or service.

Advertisers who make use of cost effective means of promotion, i.e. Internet, are advised to make maximum use of emoticons for the promotion of their product or brand as emoticons are said to spread awareness and gen y feel associated with their usage. Advertisers should make use of emoticons in corelation with social causes and awareness issues, so as to increase the awareness and attract more customers towards the product because product advertised along with social issues can leave a deep impact on the customers and thus helps in getting more customers for the particular product or the brand. Advertisers should make use of women friendly and girly emoticons in the message because this will attract the attention of women customers towards the product and also increases their liking towards the advertised message. It is recommended for the advertisers to make use of catchy, familiar and new emoticons so as to make them happy and to increase the feeling of happiness in the respondents. Therefore, correct and appropriate information about the particular product or brand should be provided by advertisers by selecting the best emoticon which can be suited well with the products and thus making the customers to purchase the product.

References

1. Alba, J., Lynch, J., Weitz, B., Janiszewski, C., Lutz, R., Sawyer, A., Wood, S.: Interactive home shopping: consumer, retailer, and manufacturer incentives to participate in electronic marketplaces. J. Mark. **61**(3), 38–53 (1997)

2. Baraniuk, C.: The surprising power of emoticons, *Future* (26 Nov 2014). Retrieved from http://www.bbc.com/future/story/20141126-surprising-power-of-emoticons
3. Bicklell, K.: Are emoticons a woman thing? *Herizons* (2014), Retrieved from http://www.herizons.ca/node/554
4. Bindu, M.H., Gupta, P., Tiwary, U.S.: Cognitive model—based emotion recognition from facial expressions for live human computer interaction. In: IEEE Symposium on Computational Intelligence in Image and Signal Processing, CIISP, pp. 351–356 (2007)
5. Consoli, D.: Emotions that influence purchase decisions and their electronic processing. Ann. Univ. Apulensis Ser. Oeconomica **11**(2), 996–1008 (2009)
6. Cristescu, I.: Emotions in human-computer interaction: the role of non-verbal behavior in interactive systems. Rev. Informatica Economica **2**(46), 110–116 (2008)
7. Crystal, D.: Language and the Internet. Cambridge University Press, Cambridge (2001)
8. DNA.: *89% of youth research online before shopping: Paradox panel report* (2014). Retrieved from http://www.dnaindia.com/money/report-89-of-youth-research-online-before-shopping-paradox-panel-report-1965642
9. Dresner, E., Herring, S.C.: Functions of the Non-Verbal in CMC: Emoticons and Illocutionary Force, *Communication Theory*, in press (2010)
10. Gharibi, S., Danesh, S.Y.S., Shahrodi, K.: Explain the effectiveness of advertising using the Aida model. Interdisc. J. Contemp. Res. Bus. **4**(2), 926–940 (2012)
11. Herring, S.C.: Gender and power in online communication. In: Holmes, J., Meyerhoff, M. (eds.), The Handbook of Language and Gender, pp. 202–228 (2003)
12. Herring, S.: Gender differences in computer mediated communication: bringing familiar baggage to the new frontier. *Miami: American Library Association Annual Convention* (1994). Retrieved from http://cpsr.org/cpsr/gender/herring.txt
13. Ho, E., Vathanophas, V.: Relating personality traits and prior knowledge to focus group process and outcome: an exploratory research. In: 7th Pacific Asia Conference on Information Systems. Adelaide, South Australia (2003)
14. How to engage millennials in customer insight and marketing.: *Vision Critical* (12 Mar 2014). Retrieved from https://www.visioncritical.com/how-engage-millennials-customer-insight-and-marketing/
15. Hwang, H.S.: Gender differences in emoticon use on mobile text messaging: evidence from a Korean sample. International Journal of Journalism Mass Communication **1**, no 107 (2014). doi:http://dx.doi.org/10.15344/2349-2635/2014/107
16. Ip, A.: The impact of emoticons on affect interpretation in instant messaging (2002). Retrieved from http://amysmile.com/pastprj/emoticonpaper.pdf
17. Kelly, J.R., Hutson-Comeaux, S.L.: Gender emotion stereotypes are context specific. Sex Roles **40**, 107–120 (1999)
18. McDuff, D., Kaliouby, R.E., Kodra, E., Larguinet, L.: Do emotions in advertising drive sales? Use of facial coding to understand the relationship between ads and sales effectiveness, Congress 2013 (2013). Retrieved from http://web.media.mit.edu/~djmcduff/assets/publications/McDuff_2013_Emotions.pdf
19. Murray, P.N.: How emotions influence what we buy: The emotional core of consumer decision making, *Inside the consumer mind* (26 Feb 2013). Retrieved from https://www.psychologytoday.com/blog/inside-the-consumer-mind/201302/how-emotions-influence-what-we-buy
20. Park, J.: Interpersonal and affective communication in synchronous online discourse. Library Q. **77**, 133–155 (2007)
21. Ptazynski, M., Rzepka, R., Araki, K., Momouchi, Y.: Research on emoticons: review of the field and proposal of research framework. In: Proceedings of the Seventeenth Annual Meeting of the Association for Natural Language Processing (NLP-2011), Organized Session on Un-Natural Language Processing, pp. 1159–1162. Toyohashi, Japan (2011)
22. Seiter, C.: The science of emotion in marketing: how our brains decide what to share and whom to trust, *Buffer Social* (04 Mar 2014). Retrieved from https://blog.bufferapp.com/science-of-emotion-in-marketing

23. Tossell, C.C., Kortum, P., Shepard, C., Walkow, L.H.B., Rahmati, A., Zhong, L.: A longitudinal study of emoticon use in text messaging from smartphones. Comput. Hum. Behav. **28**, 659–663 (2012)
24. Walther, J.B.: Interpersonal effects in computer-mediated interaction: a relational perspective. Commun. Res. **19**, 52–90 (1992)
25. Wolf, A.: Emotional expression online: gender differences in emoticon use. Cyber Psychol. Behav. **3**(5), 827–833 (2000)
26. Xu, L., Yi, C., Xu, Y.: Emotional expression online: the impact of task, relationship and personality perception on emoticon usage in instant messenger. In: 11th Pacific-Asia Conference on Information Systems, Article No. 81 (2007). Retrieved from http://www.pacis-net.org/file/2007/1267.pdf
27. Zajonc, R.B.: Feeling and thinking: preferences need no inferences. Am. Psychol. **35**(2), 151 (1980)

Analysis of Functional Parameters to Implement Knowledge Management for Sustainable e-Governance in Agriculture Sector of Saurashtra Region of Gujarat State

Alpana Upadhyay and C.K. Kumbharana

Abstract Knowledge can refer to a theoretical as well as practical or realistic understanding of a subject. Knowledge is a human sense resulting from interpreted information, an understanding that sprouts from the mixture of data, individual interpretation, information, and experience. Knowledge management is the discipline to be used to identify, capture, evaluate, retrieve, and sharing all of information resources. These resources may include procedures, documents, databases, policies, previously un-captured experiences, and expertise lying in an individual. Knowledge management offered various benefits to communities of practice, to individual employees, and to the organization itself. As we have entered in the new millennium, still the backbone of Indian economy is agriculture. In fact, the whole economy of India is being sustained by agriculture, which is also the stronghold of Indian villages. The thrust areas for the agriculture sector those needs government initiatives are availability of more and effective information, integration of entire agricultural population, train the farmers to use natural resources proficiently, equip the farmers with the latest available technologies, various cultivation processes, knowledge of crop design, various irrigation techniques, new concepts in fertilizers, pest control, timely and adequate availability of inputs, support for marketing infrastructure. For agriculture sector, the respondents are farmers of various villages of the Saurashtra region of Gujarat state. The researcher has demeanor an in-depth discussion, analysis, and testing to corroborate or reject the research findings. This helps to authenticate the significance of the outcomes and research findings.

Keywords Knowledge management · e-Governance · Agriculture · Sustainable

A. Upadhyay (✉)
Faculty of MCA, Sunshine Group of Institutions, Rajkot, Gujarat, India
e-mail: alpana.upadhyay@yahoo.com

C.K. Kumbharana
Department of Computer Science, Saurashtra University, Rajkot, Gujarat, India
e-mail: ckkumbharana@yahoo.com

A.K. Saini et al. (eds.), *ICT Based Innovations*, Advances in Intelligent Systems and Computing 653, https://doi.org/10.1007/978-981-10-6602-3_23

1 Introduction

The concept of knowledge is considered as familiarity or consciousness of somebody or something, like information, facts, skills, or descriptions that can be attained by experience or through learning, perceiving, or discovering the facts and findings. Knowledge can be referred to theoretical as well as matter-of-fact or realistic practical understanding of a given subject. Human sense and talent ensuing from human understanding, information interpretation, indulgent which takes root from various sources like experience, information, and individual construal [1].

Nonaka and Takeuchi (1995) defined knowledge as an "acceptable true belief that adds to an entity's ability for effectual action" [1]. In general, knowledge is considered as a human sense obtained from construed information, an understanding which sprouts from the mixture of data, individual interpretation, and experience. Knowledge can be defined as an "expertise; skills acquired by a person through experience or education which is the theoretical or practical understanding of a subject." It can also be "what is known in a particular field, facts and information, the awareness or familiarity gained by experience of a fact or situation." It gives the capability to make efficient decisions and take effective actions to complete the task [2].

1.1 Knowledge Management

Knowledge management has incorporated two things: (1) knowledge capturing (2) knowledge storing with the perception of preserving intellectual assets. Actually knowledge management is the purposeful and methodical and systematic synchronization of an organization's technology, people, organizational structure, and processes to put in worth by using reusability and innovations [3]. This can be accomplished by the endorsement for creating knowledge, sharing it, and finally applying it, as well as by the nourishing of the best practices and the precious lessons learned into corporate memory in order to promote continued organizational learning. There is the most excellent way to preserve valuable and important knowledge is to discover intellectual assets. Then after to ensure legacy, resources and materials are created and stored subsequently in a way that reuse and future retrieval of it is as easy as possible [4]. These tangible by products are needed to surge from one individual to other individual.

1.2 Importance of Knowledge Management in Today's Life

Today's application of KM and increased interest lie in the four key aspects [5],

1. Globalization of business: Today's organizations are global with multilingual, multisite, and multicultural in nature.

2. Leaner organizations: People do more work in a faster way, but people also require smart working as knowledge workers are—Low speed and have increased workload.
3. Corporate memory loss: People are more peripatetic as a workforce, who generates troubles of knowledge stability and continuity and it forces the load of incessant learning on the knowledge workers.
4. Advancement in technologies: People are more connected. Information Technology has promoted connectivity not only ever present but thoroughly altered prospects. The rotation time to take action is now not in weeks but in minutes.

2 Data Collection for Effective Knowledge Management for Sustainable e-Governance in Agriculture Sector

The most significant and important role in a research is of "Data Collection". Researcher can opt for the best research design but if cannot accumulate required data then would not be able to comprehend the research work. Data collection is a one of the very difficult jobs which needs to have patience, proper planning, and perseverance. It must be able to comply with the task very successfully. A popular mean of collecting data is questionnaires.

2.1 Format of Questionnaire Used for Agriculture Sector

As we have entered in the new millennium, still the backbone of Indian economy is agriculture. In fact, the whole economy of India is being sustained by agriculture, which is also the stronghold of Indian villages. Everyone among us looks toward agriculture for our nutrition too, not only economy. Rural economies are keystones of food security. Accordingly rural and agricultural problems are well thought-out to be almost identical and it is frequently understood that the objectives for both rural and agricultural can be tracked by the identical set of policies [6]. The thrust areas for the agriculture sector those needs government initiatives are availability of more and effective information, integration of entire agricultural population, train the farmers to use natural resources proficiently, equip the farmers with the latest available technologies, various cultivation processes, knowledge of crop design, various irrigation techniques, new concepts in fertilizers, pest control, timely and adequate availability of inputs, support for marketing infrastructure [7].

Following questioner was used to collect necessary data.

Dear Sir/Madam,

This questionnaire is a part of the research work and shall be used for only research purpose.

Name :______________________________
Profession :Farming______________________
Village :______________________________
Signature :______________________________

Please put a ☑ mark in the appropriate box wherever required.

The degree of your agreement, the score will be assigned on the basis of the scale, 5: Strongly Agree, 4: Agree, 3: Neutral, 2: Disagree, 1: Strongly Disagree.

In advance, I thank you and appreciate your efforts and patience to complete this questionnaire.

3 Data Projection and Analysis for Agriculture Sector

The result of the questionnaire circulated among the farmers and the responses therein have been collected. The collected responses have been projected here for the analysis one by one. The sample size has been 150. Table 1 demonstrates summary of the total responses collected for the need of knowledge management in the area of agriculture for sustainable e-Governance.

Collected data has been analyzed one by one and various charts are formed to represent that analysis in the graphical format. Following is the graphical representation of the data collected for agriculture sector. Data has been analyzed and represented as follows. Table 2 illustrates total responses on functional parameters of knowledge management of agriculture sector.

From the chart, demonstrated in Fig. 1 tells many facts related to the agriculture sector of rural areas of our country. It also demonstrates the particulars related to the need of effective knowledge management in the agriculture sector of rural India. According to the chart, more than 84% people feel that there is a slow-down in agricultural growth in rural India. 80% India is having a massive amount of population of more than one billion. It is also growing at a very high rate. This has created great demand for land. 85% people feel that automation of farming has occurred in few parts of the country, among all, greatest number of the farmers do not have enough resources to purchase modern farming tools. This obstructs the growth of agriculture. This obstructs the development of agriculture. 84% people think that to integrate the entire agricultural population in rural India flawlessly by using convergent technology solutions and promoting not just data sharing but also an information and knowledge sharing environment helps agricultural growth. 80% people strongly believe that to equip the farmers with latest available technologies and information management helps farmers in crop design. 85% people feel that the latest knowledge management makes the farmers aware about weather information. 76% people believe that use of ICT can help to boost agricultural growth. 84%

Table 1 Summary of the responses collected for agriculture sector

No.	Question	Strongly agree	Agree	Neutral	Disagree	Strongly disagree
1	There is a slow-down in agricultural growth in rural India	112	14	9	8	7
2	India is having a massive amount of population of more than one billion. It is also growing at a very high rate. This has created great demand for land	90	30	11	10	9
3	The infrastructure facilities like electricity, roads for transportation, availability of pesticides, and fertilizer are not sufficiently available in rural India that has threatened the growth of agriculture	99	30	12	3	6
4	Automation of farming has occurred in few parts of the country, among all, greatest number of the farmers do not have enough resources to purchase modern farming tools. This obstructs the growth of agriculture	114	14	10	7	5
5	To integrate the entire agricultural population in rural India flawlessly by using convergent technology solutions and promoting not just data sharing but also an information and knowledge sharing environment helps agricultural growth	96	31	9	8	6
6	To equip the farmers with latest available technologies and information management helps farmers in crop design	104	17	18	8	3
7	Latest knowledge management makes the farmers aware about weather information	106	22	14	5	3
8	Use of ICT can help to boost agricultural growth.	99	16	15	10	10
9	To make available more and effective information can support farmers in decision-making	92	34	13	9	2
10	To make available more and effective information can reduce the uncertainty and unmeasured unpredictability in agronomic conditions	93	32	9	11	5
11	Training the farmers through knowledge management can help the farmers to use natural resources proficiently	123	6	8	5	8

(continued)

Table 1 (continued)

No.	Question	Strongly agree	Agree	Neutral	Disagree	Strongly disagree
12	Effective knowledge management supports the farmers in reducing waste	115	13	5	7	10
13	Through availability of efficient information, farmers can learn from others' experiences and mistakes, thereby not repeating it next time year after year	124	0	5	17	4
14	Farmers can get sustainable raise in productivity by opting sharing of knowledge and inventive financial way outs	90	39	10	8	3
15	Using knowledge management farmers can be aware about a range of monetary options available to them to make them afford it in terms of cultivation processed	90	39	10	8	3
16	Effective knowledge management helps farmers in developing new concepts in fertilizers, pest control and farming, etc., e.g., bio-farming	111	25	6	3	5
17	Through availability of efficient information, farmers can get timely and adequate availability of inputs	119	12	6	5	8
18	Availability of adequate information supports farmers for marketing infrastructure	115	16	8	3	8
	Total	1892	390	178	135	105
	Average percentage (%)	70	14	7	5	4

people believe that to make available more and effective information can support farmers in decision-making. 83% people think that to make available more and effective information can reduce the uncertainty and unmeasured unpredictability in agronomic conditions. 86% feel that training the farmers through knowledge management can help the farmers to use natural resources proficiently. 85% people say that effective knowledge management supports the farmers in reducing waste. 82% people tell that through availability of efficient information, farmers can learn from others' experiences and mistakes, thereby not repeating it next time year after year. 86% people experience that farmers can get sustainable raise in productivity by opting sharing of knowledge and inventive financial way outs. 86% people sense that using knowledge management, farmers can be aware about a range of monetary options available to them to make them afford it in terms of cultivation processed. 90% people consider that effective knowledge management helps farmers in developing new concepts in fertilizers, pest control, farming, etc., e.g., bio-farming.

Table 2 Total responses on functional parameters of knowledge management of agriculture sector

	Q1	Q2	Q3	Q4	Q5	Q6	Q7	Q8	Q9	Q10	Q11	Q12	Q13	Q14	Q15	Q16	Q17	Q18	Total
Strongly agree	112	90	99	114	96	104	106	99	92	93	123	115	124	90	90	111	119	115	1892
Agree	14	30	30	14	31	17	22	16	34	32	6	13	0	39	39	25	12	16	390
Neutral	9	11	12	10	9	18	14	15	13	9	8	5	5	10	10	6	6	8	178
Disagree	8	10	3	7	8	8	5	10	9	11	5	7	17	8	8	3	5	3	135
Strongly disagree	7	9	6	5	6	3	3	10	2	5	8	10	4	3	3	5	8	8	105
Total	150	150	150	150	150	150	150	150	150	150	150	150	150	150	150	150	150	150	2700

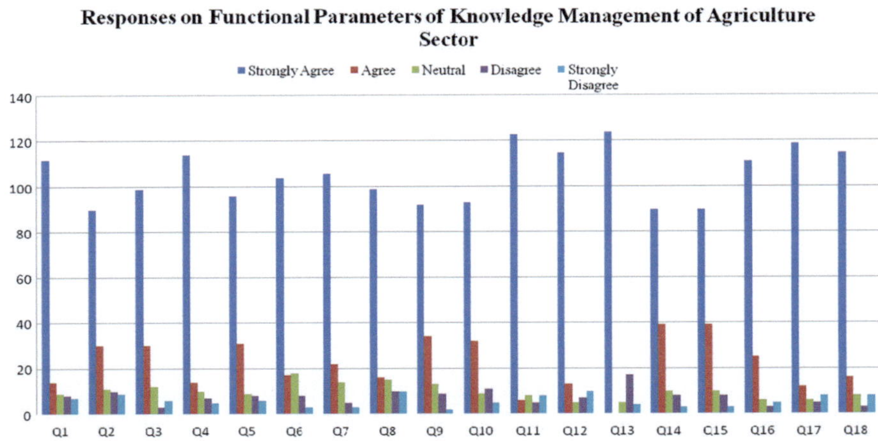

Fig. 1 Chart-1 for the responses on functional parameters of knowledge management of agriculture sector

87% people feel that through availability of efficient information, farmers can get timely and adequate availability of inputs.

- **Chart-1 for the Responses on Functional Parameters of Knowledge Management of Agriculture Sector**

There are five criteria of the responses available for data collection. To perform analysis on the basis of various criteria of the responses, researcher has selected nonparametric χ^2 test.

Research Hypothesis: Knowledge management is essential for sustainable e-Governance in agriculture sector of rural India (Table 3).

On the basis of research hypothesis, null hypothesis and alternative hypothesis are formed and tested.

H_o: Knowledge management is not essential for sustainable e-Governance in agriculture sector of rural India.

Table 3 Statistical table of nonparametric χ^2 test for agriculture sector

Criteria of the responses	Observed frequencies O_i	Expected frequencies E_i (3300*1/5)	$O_i - E_i$	$(O_i - E_i)^2$	E_i
Strongly agree	1892	660	1232	1,517,824	2299.73
Agree	390	660	−270	72,990	110.45
Neutral	178	660	−482	232,324	352.00
Disagree	135	660	−525	275,625	417.61
Strongly disagree	105	660	−555	308,025	466.70
	3300	3300			χ^2 = 3646.49

H$_a$: Knowledge management is essential for sustainable e-Governance in agriculture sector of rural India.

Statistical Data:

Level of significance = 5% => $\alpha = 0.05$

Degrees of freedom = 5–1 = 4

Critical value of χ^2 = p (Any Response) = 1/5

Thus, $\chi^2_{\text{Calculated}}$ (3646.49) > χ^2_{Critical} (9.488)

So, we reject the H$_0$ and conclude that knowledge management is essential for sustainable e-Governance in agriculture sector of rural India.

- **Chart-2 for Analysis of Total Responses Received on Various Functional Parameters of Knowledge Management for Agriculture Sector**

We can conclude from Fig. 2 that most participants responded positively, 70% strongly recommend the implementation of knowledge management for reducing time and cost of operation as well as strengthens institutional capacity in planning and managing agriculture for rural development. 14% agreed and they think it reduces time but some time it takes time to first understand the process which is neither taught nor known to them. This has to be learned during work. 7% have the neutral behavior towards the situation. They neither have much positive approach towards the change nor are they affected by the changes. 5% people are not agreed and 4% people strongly disagree to adopt any kind of changes in the system. From the feedback obtained from respondents, it is observed that 84% respondents are aware and they are strongly in favor of implementing effective knowledge

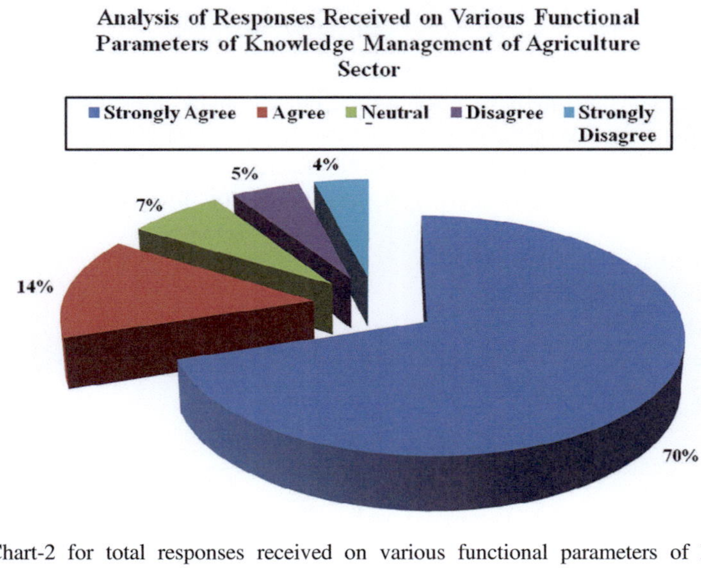

Fig. 2 Chart-2 for total responses received on various functional parameters of knowledge management for agriculture sector

management for their betterment but 16% are not aware about that and not in the favor of the same.

Research Hypothesis: Knowledge management is essential for sustainable e-Governance in agriculture sector of rural India.

On the basis of research hypothesis, null hypothesis and alternative hypothesis are formed and tested.

H_o: Knowledge management is not essential for sustainable e-Governance in agriculture sector of rural India.

H_a: Knowledge management is essential for sustainable e-Governance in agriculture sector of rural India.

Statistical Data:

Sample size = n = 150

Total respondents are 150. Total 84% respondents have given positive responses.

Hence, Number of people who agreed = 126 = x

Number of people who disagreed = 24

$$\hat{p} = \frac{x}{n} = \frac{126}{150} = 0.84 \quad \text{(Sample Proportion)}$$

$$P = 0.5 \quad \text{(Population Proportion)}$$

Since we have a sample of 150 (i.e., a large sample) and the responses can easily be classified into two categories “Agree (Believe)” and “Don’t Agree (Don’t Believe)”, we can use the *z*-test of single proportion where

$$z = \frac{\hat{p} - p}{\sqrt{\frac{p(1-p)}{n}}}$$

Since the respondents have filled up the questionnaire after thoroughly understanding, the objective of the research and the functional parameters covered in the questionnaires, we can take 99% level of confidence on the entire research process.

Hence, level of significance for the testing of hypothesis is 100–99 = 1% => $\alpha = 0.01$

H_o: P = 0.5

H_a: p > 0.5 (one tailed *z*-test)

$\alpha = 0.01$

Degree of freedom = ∞ (large sample > 100)

Critical value of ‘z’ = 2.326

$$Z\,\text{calculated} = \frac{\hat{p} - p}{\sqrt{\frac{p(1-p)}{n}}} = \frac{0.84 - 0.5}{\sqrt{\frac{0.5(0.5)}{150}}} = \frac{0.34}{0.04082} = 8.3292$$

Therefore, Z is calculated = 8.3292

$$Z_{\text{calculated}}(8.3292) > Z_{\text{critical}}(2.326)$$

Hence, H_o is REJECTED.

It is concluded that the greater proportion of the respondents "Believe" that knowledge management is essential for sustainable e-Governance in agriculture sector of rural India. So these facts itself tell the importance of effective knowledge management in the field of agriculture. It is indeed necessary to implement effective knowledge management for agriculture sector.

4 Conclusion and Key Findings

This paper provides delineate to the research methodology and rationalize the assortment of the research method and strategy. The goal of this paper is to endow with a suitable research method with the perception of the need of implementing knowledge management in various sectors of rural India for sustainable e-Governance and hence developing an adoption of knowledge management framework through the study of two major sectors of rural India. This research is concentrated on the area of e-Governance adoption in terms of knowledge management in rural India is not studied adequately and it also consents to exploration of organizational, technical and environmental barriers. The researcher has selected qualitative approach to facilitate understanding of e-Governance in the context of rural India by using interviews through questionnaire as the prime source of data for allowing in-depth examination.

It is firmly concluded that knowledge management is essential to be able to cope with societal developments and demands of rural sectors of India, to prepare them more result oriented, the process which in turn has determined the adoptability and sustainability of e-Governance. From the findings and results, it is strongly proposed to develop a conceptual knowledge management framework for the implementation of new structures and strategies.

References

1. Locke, J.; Nonaka & Takeuchi and Andre Boudreau. Accessed 01 June 2015 http://www.businessdictionary.com/definition/knowledge.html
2. "The Prospect of Knowledge Creation-Laurea SIDLabs", Frank Nyarko (2009)
3. "Knowledge-Based Systems", Jones & Bartlett publishers, Canada (2010)
4. Stewart, T.: Software preserves knowledge, people pass it on. Fortune **142**(5), 4 (2000)
5. Koenig Michael, E.D.: "What is KM? Knowledge Management Explained", KM World Magazine (2012)

6. Upadhyay, A., Kumbharana, C.K.: "Development of knowledge based framework for agriculture sector: a step towards sustainable e-governance in Rural India". Int. J. Res. Comput. Application Manage. **3**(02) (2013) (February). ISSN 2231-1009
7. The Agriculture Ministry, "The State of Indian Agriculture 2011–2012", Posted by Target UPSC on March 30, 2012, Available at: http://yourage.wordpress.com/2012/03/30/the-state-of-indian-agriculture-2011-2012/, Accessed 10 May 2015

Trust and Distrust in Web: Two Sides of a Same Coin or Poles Apart?

Himani Bansal and Shruti Kohli

Abstract Both computer and social researchers have done an incredible work in identifying factors influencing trust. Though social researchers limit their findings to trust between relationships, computer people extend these factors and many more to estimate trust in the world of web. While working on trust as a subject, distrust often tends to be forgotten as very limited work is encountered in this area. This paper is an attempt to study and distinguish distrust in various forms of web especially social networking and websites. A survey is used to ascertain the results to differentiate between these two web media.

Keywords Trust · Distrust · Web trust · Social trust · Distrust in web

1 Introduction

Trust is defined as "reliance on the integrity, strength, ability, surety, etc., of a person or thing", whereas distrust is defined as "to regard with doubt or suspicion; have no trust in or lack of trust" [1]. They are reciprocal to each other or considered as two faces of a same coin. But these definitions are in context of social relationships. In present times of digital world, when people rely heavily on internet for information, advice, relationships counseling, education, social networking, etc., both trust and distrust become a vital factor to analyze the credibility of these sources. The authors have already used behavior modeling of visitors to establish trust value of content-driven websites [2–5]. To add to it, it was further decided to study about the distrust value possessed by these websites. For initiating this study, a survey was conducted among the Internet users to identify the behavioral highlights of the users of the websites. The questionnaire circulated consisted of nine

H. Bansal (✉) · S. Kohli
Computer Science Department, BIT Mesra, Ranchi, India
e-mail: singal.himani@gmail.com

S. Kohli
e-mail: kohli.shruti@gmail.com

A.K. Saini et al. (eds.), *ICT Based Innovations*, Advances in Intelligent Systems and Computing 653, https://doi.org/10.1007/978-981-10-6602-3_24

questions based on the selected factors as detailed in [6]. After formulating the questionnaire, we decided to use convenience sampling of (minimum) size 100. We collected the responses until we reach this number. In all, we collected 116 responses. All of these passed the validation phase. To ascertain single response from a single respondent, appropriate validation check was implemented.

2 Survey and Results

The first two questions of the survey were background questions inquiring about gender and age of the respondents. Figures 1 and 2 depict the responses collected in a consolidated visual manner. This information is collected to further analyze the dependency of behavior of website visitors with gender and age.

The third question posed inquiry as to whether visitors turn up to Internet for gathering information, to which 115 respondents out of 116 gave affirmation; thus again strengthening the fact that Internet acts as a mass media for information gathering and hence a method for investigating trust or distrust is highly advisable (Fig. 3).

Questions 4–7 in the survey were scaled according to the Likert scale with range varying from 1 to 5, where 1 points completely false statement and 5 points to completely true statement. These questions were framed to study the relationship between identified web metrics from previous studies [2, 6] and distrust generated. Figure 4 helps in analyzing whether visitors distrust any information in websites. Major respondents' response lies between 2 and 3 which suggests that website users tend to believe or trust the information in websites but have a little amount of distrust also which should be further analyzed in detail.

Figure 5 depicts the relationship of website quality with distrust. It can be observed that as the website quality decreases, people tend to distrust the information but it is not a clear indicator of distrusting any information.

Fig. 1 Gender distribution

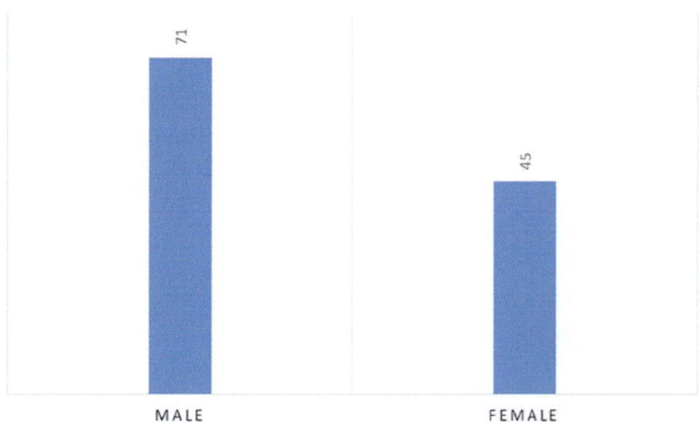

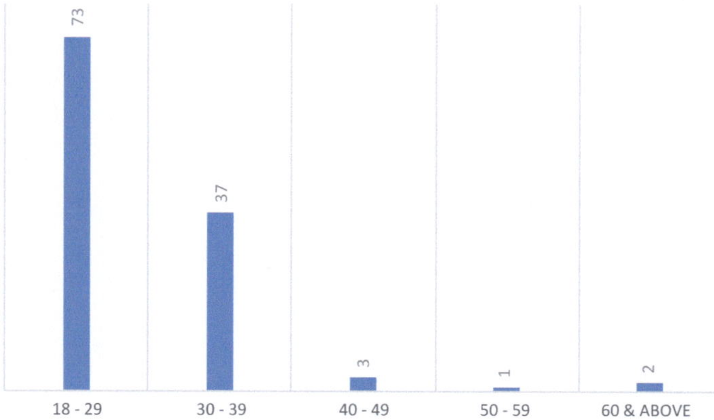

Fig. 2 Age distribution

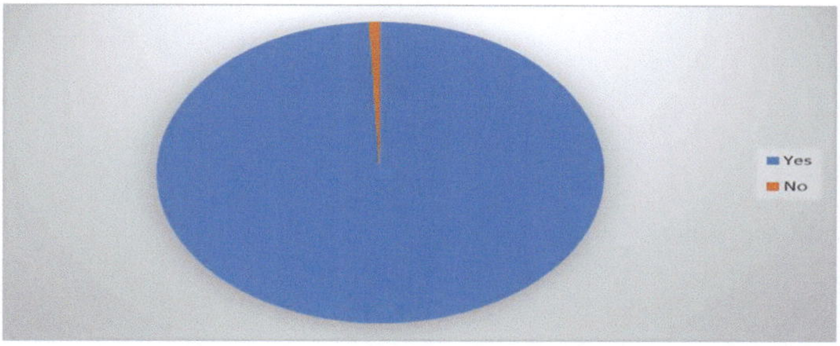

Fig. 3 Usage of Internet for gathering information

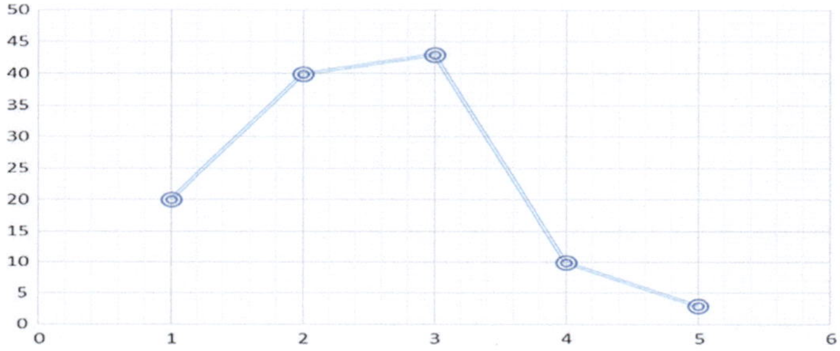

Fig. 4 Tendency of distrusting information from websites

In a similar way, Figs. 6 and 7 help in establishing relationship of distrusting information with returning visitors and bounce rate, respectively. It is observed that both these factors/metrics have a direct relationship with behavior of distrusting information. If the visitors do not like the information or distrust the information, they do not return to the website and bounce rate of the website increases.

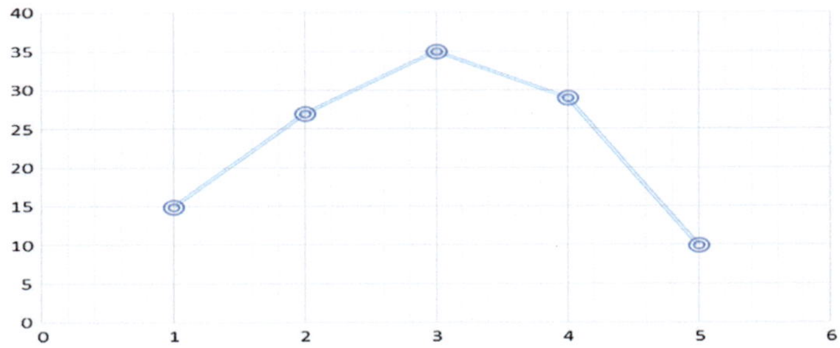

Fig. 5 Distrust and website quality

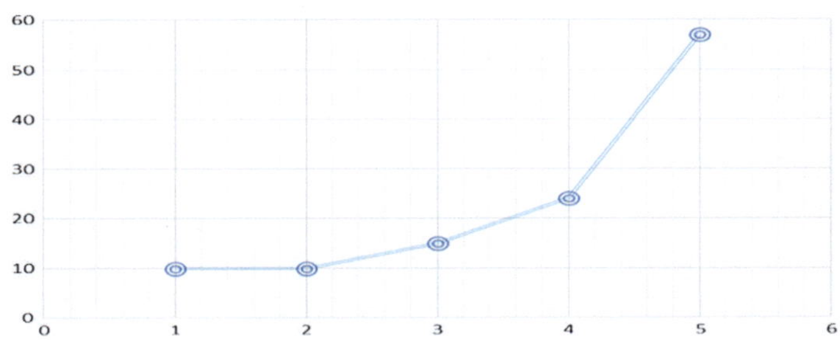

Fig. 6 Distrust and visitors

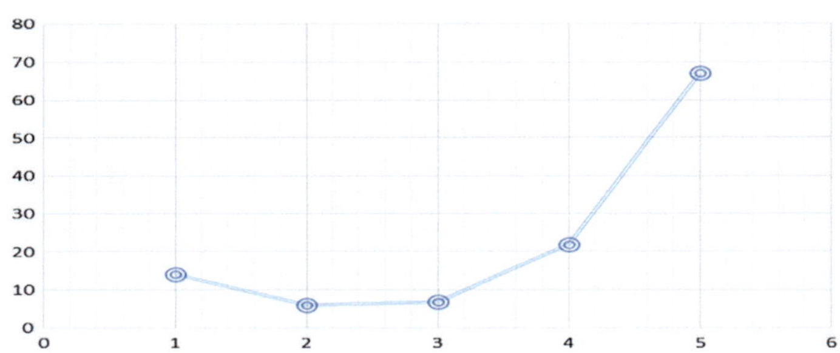

Fig. 7 Distrust and bounce rate

Figures 8 and 9 show the relationship between distrust and number of visitors of the websites and revisits made by the visitors. Majority of the respondents affirmed that if they distrust the information, they will not visit the website again, whereas number of visitor count/hit counter has somewhat neutral effect, but major portion inclining towards a relationship that if hit counter is low, then people somewhat tend to distrust the information.

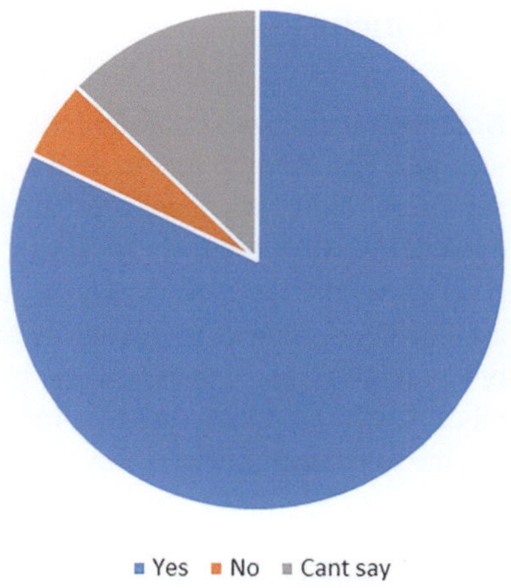

Fig. 8 Distrust and revisit by the website user

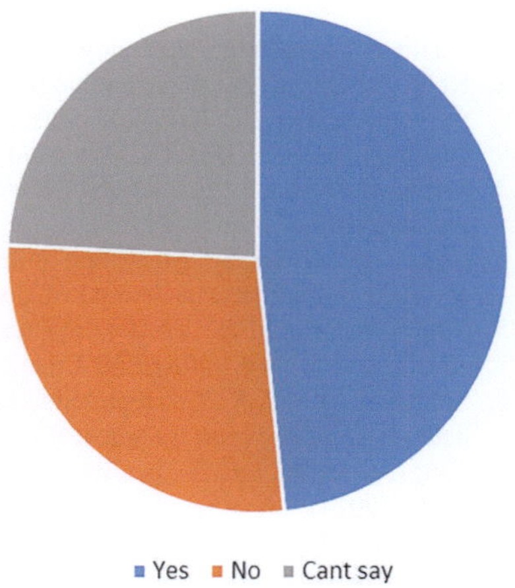

Fig. 9 Distrust and visit counter

3 Discussion

The survey was designed to study the relationship between the specific web metrics (Average time on website, Pages/Visit, Average Daily Visits and Bounce Rate) and distrust in information provided by websites to enhance the study of authors about quantification of trust and distrust value of the websites. Other questions were framed to support these theories and maintaining continuity in the survey. The analysis of the responses tracked by this survey suggested a complimentary relationship between trust and distrust, i.e., they are reciprocals of each other. This opposes the power law like function and balance theory introduced by many researchers [7, 8]. This is majorly because all these studies are done in the area of social media websites where concept of transitivity exists between both the trust value and distrust value generated, progressed and then finally consumed.

4 Conclusion

Both trust and distrust are ubiquitous and their influence can be felt in the world of web too. The drive to find trust in web is essential to provide credible information to the users. With trust, concept of distrust also incorporates, so its study is also essential parallel to the study of trust. This is because they are believed to follow the power law like function. Through our survey, we concluded that power law is only applicable in context of social media or social linking websites. For informational or content-driven websites, both trust and distrust exist in reciprocation of each other. It can be concluded that for content-driven websites, both trust and distrust are the two faces of the same coin.

References

1. Dictionary.com: http://dictionary.reference.com/ (accessed on 14/09/2015)
2. Kohli, S., Singal, H.: A methodological approach for Quantizing Trust from Human Behavior for Content-Driven Websites: Special Issue, Cyber Security, Privacy and Forensics for Complex Systems, IJSWCC, Bentham Publishers (2015)
3. Singal, H., Kohli, S.: Escalation of Trust Analysis in Web: 12th ACM International Conference on Computing Frontiers (2015)
4. Singal, H., Kohli, S.: Mitigating Information Trust—Taking the edge off health websites: accepted in International Journal of Technoethics, IGI Global
5. Singal, H., Kohli, S.: Trust necessitated through metrics: estimating the trustworthiness of websites: Accepted in Procedia Computer Science, Elseveir
6. Singal, H., Kohli, S.: Conceptual model for obfuscated TRUST induced from Web Analytics data for content-driven websites: IEEE ICACCI (2014)
7. Tang, J., Hu, X., Liu, H.: Is distrust negation of trust? ACM HT, The value of distrust in social media (2014)
8. Tang, J., Hu, X., Chang, Y., Liu, H.: Predictability of distrust with Interaction Data: ACM CIKM (2014)

Computer Simulation Using GPSC Package MATLAB, Simulink for Bioinformatics Professional

Kiran Nehra, Vijay Nehra, Bhupinder Singh, Sunil Kumar and Mahesh Kumar

Abstract Scientific computing tool encompasses vital role in science and engineering education. General purpose scientific computing (GPSC) tool MATLAB, Simulink, its toolboxes and block sets are well known as leading simulation package across globe. It is a standard de facto simulation tool across multiple disciplines of science and engineering education and research. In this investigation, the usage of GPSC tool is demonstrated in educational effectiveness in bioinformatics programme. The major goal of such education approach is to produce professionally qualified engineers who are skilled in the usage of the same and are ready for innovation in research and industry careers, able to share result, and collaborate with international projects.

Keywords Scientific computing · MATLAB · Simulink · Bioinformatics · Biotechnology

---

K. Nehra (✉)
Department of Biotechnology, Deenbandh ChhotuRam University of Science and Technology, Murthal, Sonepat, Haryana, India
e-mail: nehrakiran@gmail.com

V. Nehra
Department of Electronics and Communication Engineering, Bhagat Phool Singh Mahila Vishwavidayalaya, Khanpur Kalan, Sonepat, Haryana, India
e-mail: nehra_vijay@yahoo.com

B. Singh · S. Kumar
Department of Basic and Applied Sciences, Bhagat Phool Singh Mahila Vishwavidayalaya, Khanpur Kalan, Sonepat, Haryana, India

M. Kumar
Department of Electronics and Communication Engineering, Shree Mata Vaishno Devi University, Jammu, Jammu and Kashmir, India

A.K. Saini et al. (eds.), *ICT Based Innovations*, Advances in Intelligent Systems and Computing 653, https://doi.org/10.1007/978-981-10-6602-3_25

1 Introduction

The economic development of any nation depends on human resource development, particularly in science and engineering as well as technological advancements and industrialization. Presently, science and engineering education horizon is expanding nationwide, cultivating innovative talents across science and engineering education is the need of hours to make globally competitive professional. This is especially challenging and urgent because of the rapid growth of the knowledge-based economy and society. Student-centered active learning, autonomous learning, and hands on exposure through computer simulation technique using virtual tools MATLAB, Simulink play a pivotal role in this direction [1–22].

Looking at bioinformatics, it is a truly multidisciplinary field and requires domain knowledge of basics in biology, mathematics, and computer science [15–22]. Presently, a wide range of simulation platform and various programming language support the education and research in bioinformatics domain. Mehmood et al. presented application of bioinformatics tools in various areas of biological sciences [22]. In general, an overview of simulation environment is depicted in Fig. 1. It is well known that industry standard general purpose scientific computing (GPSC) tool.

MATAB, Simulink is assimilated as standard de facto simulation tool in multidiscipline of science and engineering domain as well as bioinformatics. However, much has not been explored in this direction in the context of life science in state of Haryana. It restricts students learning, their professional growth and leads to hindrance in nurturing application oriented professional.

In the succeeding section, motivation and rationale for assimilation of GPSC particularly in life science and in context of bioinformatics domain are presented.

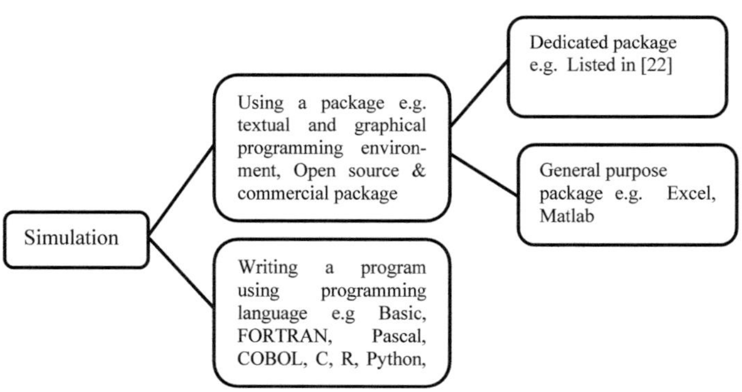

Fig. 1 Outlook of simulation options [Modified from 13]

2 Motivation and Rationale for Assimilation of GPSC

Certainly, much has not been explored in the context of assimilation of GPSC in science and particularly in basic and applied science, life science across state universities of Haryana. In general, based on discussion and interaction with academic community, some basic consideration in the context of usage of the same is presented in the forthcoming subsection.

2.1 Some Basic Considerations: Outlook

In knowledge base society, cultivation of innovation among youth is the need of hours. Presently, India has a sound network of technical and professional institutions and has a huge potential. There are several issues and challenges in the nurturing of innovation for future. Therefore, it is imperative to identify the key issues that prevent the assimilation of the same in education support. Keeping this in view, some basic considerations which need to paid attention in the context of role and usage of GPSC in biotechnology, bioinformatics, biomechanics, forensic science areas through literature survey and interactions with academic experts are:

1. In author's view, only very few human resources in "Faculty of Basic and Life Science" across state universities teaching departments are utilizing MATLAB in students-centered instruction delivery, classroom presentation, and laboratory exercise.
2. Usually, the pedagogical philosophy in basic and life science is traditional approach having a focus on theory without any practical and real life time situation, without assimilating of programming language, open source environment, and commercial simulation platform.
3. Shortage of professionally trained human resources to utilize GPSC tool and difficulty of using the same across domains in minor/major project, dissertation, and research work. However, there is major emphasis on R&D projects in basic and life science.
4. Student's awareness towards usage of MATLAB, Simulink environment, bioinformatics toolbox in their research activities and autonomous learning is negligible due to lack of proper assimilation in curricula.
5. The awareness to use bioinformatics toolbox is very limited.
6. Moreover, following challenges are also vital for cultivation of innovation:

   (a) Lack of programming, problem-solving skills, and customized solutions among students.
   (b) Lack of coordination and lack of experts at decision and policy making level in the UG and PG board of studies for assimilation of the same.
   (c) Lack of ICT infrastructure and resource across institution.

(d) Lack of centralized learning resource center for nurturing skilled personnel and training in GPSC.

To sum up, there is low motivation and less inclination of life science professionals towards the usage of GPSC across curricula and their professional career.

2.2 Assimilation Across Curricula

In author's opinion, students are generally not exposed to the scientific computing tool MATLAB, Simulink during their course of studies in the context of bioinformatics and biotechnology domain. Truly speaking, due to lack of assimilation across the domain in curricula, students are unable to translate analytical solution into simulation. Consequently, student's problem solving and simulation skill are lacking. Moreover, due to lack of assimilation, there is no exposure of MATLAB fundamental syntax, advance programming capabilities, MATLAB GUIDE feature specialized toolbox, capabilities and application, awareness of model based design using Simulink, and Simulink block sets library and creation of customized application.

3 Objective of Study

The objective of the present study is:

1. To explore assimilation of scientific computing tool MATLAB, Simulink, its toolbox as well as to understand its usage in cultivation of outstanding application oriented professional and hands on experience workforce.
2. Engaging students in learning industry standard mathematical application package MATLAB, Simulink, its toolboxes and supporting the usage of the same in independent autonomous learning across their professional career.
3. Analysis of potential versus reality in the context of assimilation from student's instructor and institutional point of view.

In the forthcoming section, an analysis of assimilation of the MATLAB, Simulink, its toolbox, blocksets, and FOSS in UG and PG bioinformatics and biotechnology curricula are presented.

4 Brief Outlook: GPSC MATLAB

GPSC MATLAB is a high-level technical computing language. It is a commercial scientific computing environment for mathematical modeling, simulation, and data analysis. It is an invaluable tool for systems applications from research level to implementation and is widely used in mathematical analysis, soft computing, automatic control engineering, image and signal processing, biomedical engineering, optimization, bioinformatics, and other fields. In international academia, it has been recognized as accurate and reliable scientific computing standard software. Simulink is a MATLAB add-on package that can be used to model and simulate dynamical process. It enables block diagram representation of system and running simulation very quickly. It is a tool of choice for control system design, digital signal processing design, communication system design, and other application. The usage of the same has increased consistently in academic community in multiple domains as well in industry [1–22]. In the forthcoming subsection, a summary of benefit of blending of GPSC tool is presented.

4.1 Role and Usage of GPSC Tool MATLAB in Biotechnology and Bioinformatics

GPSC MATLAB and add-on specific toolboxes such as bioinformatics toolbox, statistical toolbox, and simbiology toolbox enable the applicability of the same in interdisciplinary field of bioinformatics. It enables to study complex biological process and phenomena that are extremely difficult to achieve due to high cost of experiments. A framework highlighting the role and usage of the same is depicted in Table 1.

The main features of bioinformatics toolbox are [15–18]:

- Next generation sequence analysis and browser sequence analysis and visualization, including pair wise and multiple sequence alignment and peak detection
- Microarray data analysis, including reading, filtering, normalizing, and visualization
- Mass spectrometry analysis including preprocessing, classification, and marker identification
- Phylogenetic tree analysis
- Graph theory function including interaction maps, hierarchy plots, and pathways
- Data import from genomic, proteomic, and gene expression files, including SAM, FASTA, CEL, and CDF from data bases such as NCBI and Genbank.

Major features of GPSC are: high-level technical computing language; user-friendly; wide range of built in function; simpler in programming and having programming compactness; interfacing with other programming languages such as C, C ++, Java, Excel, and possibility of automated generating code from Simulink

Table 1 Role and usage of scientific computing tool MATLAB [15–22]

MathWorks product description	Usage level and skills required
MATLAB: Text programming	• MATLAB script solution across domain • Require MATLAB programming skills
Simulink: Dynamic system modeling	• Biochemical process modeling • Require Simulink library blocks awareness and graphical programming solution across domain
Data-driven modeling	• MATLAB and Simulink together • Require model based design and programming skills
Bioinformatics toolbox	• Primary research and training software for biological application • Sequence analysis • Microarray analysis • Mass spectroscopy data analysis • Graph theory application • Enable students to develop customized algorithm to investigate and understand current advancement in the area of genetic engineering and other genomic and proteomics
Neural network toolbox	• As ancillary tool, allow students and researcher to explore complex, nonlinear, multivariable functions
Statistical toolbox	• As ancillary tool, enable students to perform complex statistical tests on data quickly and efficiently
Simbiology toolbox	• As Simbiology toolbox in system biology • Modeling tools used in system biology

model. The interested reader may refer to MathWorks site for detailed information about usage, book and product description pertaining to life science [1].

4.2 Framework for Assimilation

It is obvious from literature survey and as per discussion with academic community that the assimilation of GPSC during the first year as problem-solving and programming tool play a pivotal role. The authors believe that exposure and utilization of MATLAB as problem-solving tool in almost all course at some point across the curriculum will increase students understanding and comfort level, making them highly skilled and confident users of GPSC.

4.3 Benefits of Blending Scientific Computing Tool

The benefits of blending scientific computing tool in biotechnology, bioinformatics, biomechanics, and forensic science areas are as follows:

- Provides an in-depth understanding of life science principals and concept.
- Enables student-centered active learning and autonomous learning.
- Enhances problem solving, simulation skills and employability of graduates.
- Enables to study dynamic behavior of regulatory mechanism involved in signal pathways of living system, moreover, dynamic modeling of biological and biochemical system using graphical programming environment Simulink.
- Usage as teaching and learning aids, as pedagogical tool in teaching and research in life science domain.
- Enables to supplement traditional laboratory with GPSC environment to capture and explore experimental data and as an experimentation tool.
- Analysis of biotechnology and bioinformatics laboratory data for presenting results in the form of graphs, plots, and simulation.
- Modeling and experimentation tool in conducting of laboratory exercise.
- Enables to analyze the same problem at various levels and helps students to understand the theory deeply and thoroughly.
- Enables to utilize the same in compilation of minor/major project work and dissertation later on during their course of study. It is useful for further research and professional growth. Moreover, enables to use in customized application solution.
- Accelerating innovation through simulation-based engineering and sciences.

In nutshell, it provides much more effective and efficient environment in teaching and learning thereby resulting in efficient education in terms of time and cost.

4.4 Multiple Level of Usage of Scientific Computing Tool

An advanced user can solve the real-world problems pertaining to bioinformatics, biomechanics, biotechnology, and forensic science in multiple ways by using scientific computing tool at various levels as mentioned here:

- First level: by using powerful functions directly.
- Second level: by using scientific computing tool to realize the biological process.
- Third level: by developing new optimizing algorithms.
- Fourth level: Model-based design using graphical programming tool Simulink in workflow.
- Fifth level: by using specialized add-on toolbox.

To sum up, multiple usages of the same across domain enhance problem solving, process thinking skills, and thus better prepare students to pursue professional career in interdisciplinary field and led innovation for future.

4.5 As Pedagogical Tool in Teaching and Research

The usage of MATLAB as primary resource supplements the bioinformatics and biotechnology foundation and core courses across domain. It improves students understanding of fundamental concepts, and increases students interest and performance.

4.6 Supplementing Traditional Laboratory

This section focuses on the usage of MATLAB textual programming environment, graphical user interface feature, model-based design feature using graphical environment Simulink, and a wide range of add-on toolbox across biotechnology major course laboratory curricula. There are several examples illustrating the usage of the same by various academicians as simulation tool in virtual laboratories. A lot of literature available on MathWorks and Internet supports its usage at multiple levels across curricula. Many prestigious universities have incorporated MATLAB tool for student learning. The assimilation across laboratory curricula enhances knowledge and understanding, intellectual abilities, practical skills, and transferable skills.

It enables students to use MATLAB as cheap and accessible within laboratories across generic bioinformatics and biotechnology major courses across domain curricula, thereby addressing the issue of lack of physical resources. The same augment student's hands-on exposure by using MATLAB at multiple levels without requiring them to enter a time table laboratory. It enables student's professional career development through simulation laboratory. The students can realize the laboratory experiment using built in function, translating algorithm into codes, realization using GUI feature GUIDE of MATLAB, dedicated toolbox, and experiment.

In brief, it enables to achieve the mission of skill development and make in India by cultivating outstanding application oriented and hands-on experience workforce.

5 Potential Versus Reality Across State

This section summarizes the potential versus reality about the usage of mathematical application package MATLAB, Simulink and add-on toolbox in context of effective utilization by students; assimilation across curricula and instructor point of view. An analysis of potential versus reality in the context of assimilation and usage of the same on student's outcomes are presented in Table 2.

Table 2 Potential versus reality of MATLAB in context of students

Description	Potential	Reality
Effective assimilation in curricula for enhancing deeper learning of students	Yes, enhance deeper learning across curricula	Not utilized across curricula due to lack of assimilation
Effective utilization by students in biotechnology foundation and core courses	Yes, foundation and advanced course	Not utilized, across theses levels
Effective utilization of ICT resource and simulation tool across multiple courses	Yes, address the lack of resource	Not utilized in multiple discipline
Effective utilization in autonomous learning	Yes, promotes self and independent learning	No awareness and not utilized
Effective utilization of graphical programming through Simulink	Yes, across curricula in model based design	Not utilized due to lack of awareness
Effective utilization of textual feature in laboratory exercise to capture and explore experimental data by students	Yes, but not across curricula	Not assimilated across curricula and not utilized by students
Effective utilization in minor and major project work in UG education	Yes, enable to complete the project	No awareness
Enable to achieve to enhance knowledge and understanding across curricula	Yes	Not assimilated across curricula
Computational based MATLAB exercise and project for students	Yes	No
Effective usage by students in creating cheap and accessible virtual laboratories	Yes	Potential not tapped across state
Improve students' learning problem solving in new situations	Yes	Not utilized

5.1 *Analysis of Potential Versus Reality in Context of Curricula*

Table 3 provides an outlook of potential versus reality about the usage of mathematical application package in the context of assimilation across curricula.

5.2 *Analysis of Potential Versus Reality in Context of Instructor*

An analysis of potential versus reality in the context of usage of scientific computing tool in teaching domain core courses from an instructor point of view is presented in Table 4.

Table 3 Analysis of potential versus reality in context of curricula

Description	Potential	Reality
Effective assimilation across curricula in basic science, foundation and major courses	Yes, enhance deeper learning	Not integrated across curricula
Effective assimilation of graphical programming and GUI feature	Yes, across curricula	Not integrated in curricula
Able to cultivate innovative talents and teaching reform across state	Yes	Not assimilated effectively in educational process
Cultivation of outstanding application oriented professional through assimilation across curricula	Yes	Not used by the students
Effective assimilation to enhance knowledge and understanding, intellectual abilities, practical skill, and transferable skills	Yes	Not utilized

Table 4 Potential versus reality in context of instructor

Description	Potential	Reality
Effective utilization in biotechnology curricula and problem-based learning	Offer vast potential	Used only in bioinformatics laboratory
Able to nurture innovative talents and teaching reform across state	Yes	Not utilized effectively in educational process
Blended with traditional laboratory to supplement experiment across curricula	Yes	No utilized to capture date and explore experimental
Enhance knowledge and understanding across curricula	Yes	Not utilized
Effective utilization in enhancing knowledge and understanding, intellectual abilities, practical skill and transferable skills	Yes	Not utilized
Extracting grants for organization of STTP/FDP/Workshop supporting usage of the same	Yes	Not conducted specially in domain of life science
Effective usage in creating cheap and accessible virtual laboratories	Yes	Potential not tapped

6 Conclusion

In this study, the usage of GPSC package MATLAB, Simulink, and its toolbox as support for education and research in bioinformatics is demonstrated. The bioinformatics course is a rather challenging subject for the instructor and students pursuing bachelor's and master's degree in biotechnology and bioinformatics engineering. The students can use effectively GPSC MATLAB as textual programming in multiple courses and at multiple levels such as built-in function, algorithm to develop code, dedicated toolbox, GUIDE feature, and graphical programming Simulink and blocksets across curricula if they have strong hands on experience during their course of study. The assimilation of GPSC is certainly an instrument for success of educational process. It helps educator to produce technical skilled qualified professional who is ready for innovation in research and industry career and cultivate outstanding application oriented engineer. Finally, the present communication proposes the integration of MATLAB package in UG and PG biotechnology and bioinformatics domain across the state.

References

1. The Math Works Inc., Natick, Massachusetts, MATLAB, 1993: http://www.mathworks.co.uk
2. Kalechman, M.: Practical MATLAB applications for engineers, CRC Press (2009)
3. Blaho, M., Foltin, M., Fodrek, P., and Murgas, J.: Education of Future Advanced MATLAB Users, MATLAB-A Fundamental Tool for Scientific Computing and Engineering Applications-Volume 3, InTech (2012)
4. Blaho, M., Foltin, M., Fodrek, P., Poliacik, P.: Preparing advanced Matlab users. WSEAS Transaction on Advances in Engineering Education **7**, 234–243 (2010)
5. Feng, P., Mingxiu, L., Dingyu, X., Dali, C., Jianjiang, C.: Application of Matlab in teaching reform and cultivation of innovation talents in universities, In: IEEE 2nd International Workshop on Education Technology and Computer Science, pp. 700–703 (2010)
6. Tahir, H.H., Pareja, T.F.: Matlab package and science subjects for undergraduate studies. International Journal for Cross-Disciplinary Subjects in Education **1**, 38–42 (2010)
7. Jain, S.: Modeling and simulation using Matlab-Simulink, Wiley, India (2011)
8. Dabney, J.B., Harman, T.L.: Mastering Simulink, Pearson Education (2004)
9. Karris, S.T.: Introduction to Simulink with engineering applications, Orchard publications (2006)
10. Michal A. Gray, Introduction to the simulation of dynamics using Simulink, Chapman & Hall/CRC, 2011
11. Nehra, V., Tyagi, A.: Free open source software in electronics engineering education: a survey. International Journal of Modern Education and Computer Science **6**, 15–25 (2014)
12. Nehra, V.: MATLAB/Simulink based study of different approaches using mathematical model of differential equations, I.J. Intelligent Systems and Applications, **6**, 1–24 (2014)
13. Ibrahim, D.: Engineering Simulation with MATLAB: improving teaching and learning effectiveness, In: Procedia Computer Science, 3, 853–858 (2011)
14. Ryan, J., Tarcini, A.: Preparing students to accelerate innovation through simulation based engineering and sciences, In: IEEE 2nd International Workshop proceedings on Education Technology and Computer Science, 2,700–703 (2010)
15. Bioinformatics Toolbox User's Guide, The Math Works Inc., Natick, Massachusetts, MATLAB, 1993: http://www.mathworks.co.uk
16. Bioinformatics Toolbox™, User's Guide, http://uk.mathworks.com/help/pdf_doc/bioinfo/bioinfo_ug.pdf
17. MATLAB Applications in Bioinformatics, http://www.ftp.mi.fu-berlin.de/bkeller/MatLab/Examples/Bioinformatics.ppt
18. Application of Matlab in Bioinformatics, http://www.srmuniv.ac.in/downloads/bi0500.pdf
19. Narayanan, G.: Teaching of Essential Matlab Commands in Applied Mathematics Course for Engineering Technology, http://www.icee.usm.edu/icee/conferences/asee2007/papers/3055_TEACHING_OF_ESSENTIAL_MATLAB_COMMANDS_IN.pdf

20. Sarawut Wongphayak Using Matlab: Bioinformatics Toolbox For Life Sciences, www.bioinformatics.kmutt.ac.th/download/seminar/.../Sarawut_PPT2.pdf
21. Dealing with biological data base, https://www1.ethz.ch/bsse/cbg/education/fall2013/bioinf/matlab2_DBs.pdf
22. Mehmood, M.A., Sehar, S., Ahmad, N.: Use of Bioinformatics Tools in Different Spheres of Life Sciences. J Data Mining Genomics Proteomics **5**, 1–13 (2014)

MIX
Papier aus verantwortungsvollen Quellen
Paper from responsible sources
FSC® C105338

If you have any concerns about our products,
you can contact us on
ProductSafety@springernature.com

In case Publisher is established outside the EU,
the EU authorized representative is:
Springer Nature Customer Service Center GmbH
Europaplatz 3, 69115 Heidelberg, Germany

Printed by Libri Plureos GmbH
in Hamburg, Germany